# MECHANISMS OF LANDSCAPE REHABILITATION AND SUSTAINABILITY

## Authors

### Valentin Kefeli

Slippery Rock Watershed Coalition
Pennsylvania, USA

### Winfried Blum

University of Natural Resources
Applied Life Sciences
Vienna, Austria

## Editor

### Narcin Palavan-Unsal*

Istanbul Kultur University, Istanbul, Turkey

*Address Correspondance to: Dr. Narcin Palavan-Unsal, Istanbul Kultur University, Department of Molecular Biology and Genetics, Atakoy Campus, 34156 Bakirkoy, Istanbul-Turkey, Tel: 00902124984347; E-mail: n.palavanunsal@iku.edu.tr

# eBooks End User License Agreement

# CONTENTS

## CHAPTERS

### Part I: Soil as a Component of the Biosphere

### Part II: General Characteristics for Fabricated Soil

### Part III: Biomass Accumulation and Growth Regulation Processes in Plants Growing on Fabricated Soil

### Part IV: Water Cycle and Plant Activity

# FOREWORD

An achievement of sustainability demands a better understanding of plant-soil relationships as agriculture remains a primary activity to sustain a growing human population. 'Mechanisms of Landscape Rehabilitation and Sustainability' by Valentin Kefeli and Winfried Blum (Narcin Palavan-Unsal, Editor) fulfill this objective through a comprehensive presentation, which is articulated in four parts, and twenty-two chapters (Pp. 191). Initially, the authors present soil as a component of the biosphere. Such an important resource deserves special attention as topsoil loss undermines food security, on a global scale. Thus, the need to develop more sustainable land management practices is predicated and approaches to 'construct' soil from renewable biomass becomes an interesting aspect, which is presented at the beginning of the book.

Part two elaborates a step further on practices aimed at rehabilitating soil through what the authors call "a fabricated soil'. Emphasis to this effort is given to microbial activities as bacteria and fungi play fundamental roles in sustaining life within a soil system. These biocenoses are conducive to a stabilization of the soil organic matter. As the humification process leads to the mineralization of carbon rich macromolecules, and as the cycle of plant, nutritional elements is completed, allowing nutrient uptake by the root systems of the plant community.

Part three presents plant physiological processes like growth, photosynthesis and hormones interactions and how these interact with the soil. Chapter 17 can be of particular interest to readers who wish to better understand the implications of phenomena like allelopathy in the cultivated field. The success strategy to overcome soil-sickness appears to be linked to the compelling need of increasing biodiversity in the soil and this can be achieved by diversifying the biomass employed to make a fabricated soil.

The concluding part of the volume (Part four) connects soil and plants with water and the hydrologic water cycle. Plants are at the interface between lithosphere and atmosphere and thus the stability of plant communities play a vital role for maintaining the equilibria of terrestrial ecosystems. However, plants are also employable to recycle polluted water and wetland construction and maintenance inspire readers to want to learn more about this practical application and use of plants. The conclusion section illustrates briefly the authors' breadth and depth of knowledge of plant biology, soils and applied ecology. Their effort to connect several sciences together in a language that is easily understandable should be commended. This book can be of use to the professional practitioner (e.g. ecological engineer, restorationist, landscape architect and others) and also to students of biology and life applied sciences.

**Professor Bruno Borsari**
Winona State University
USA

# EDITORIAL PREFACE

The conception of landscape rehabilitation is based on the investigation of natural processes, which proceed in the soil and in the plants. Authors of this book not only describe the soil function but also their ideas on sustainability and regeneration of the eco-sphere elements and human participation in these processes.

*Homo sapiens,* during their further evolution, need to develop both an understanding and a practical means to protect themselves against mechanisms of self-degradation. As integral parts of the Earth's ecosystem, intelligence demands that humans protect and maintain the various functions of the primary biosphere cycles so as to prevent the loss of soil and desertification, the pollution of air and water, the formation and expansion of the hole in Earth's ozone layer, the reduction of biodiversity, and global warming. The evidence is formidable that these harmful processes are the results of human activity in the industrial age.

This book presents discoveries and proposals that have emerged from the authors' research and experimentation on ways to protect Earth's ecosystems against further degradation. These proposals are founded on the philosophy of sustainable development and its application to various aspects essential to the long-term success of human beings: community coexistence, education, water purification and recycling, agriculture, the production of fabricated soil for landscape rehabilitation, and the preservation and propagation of wild flora. Strategies for both indoor and outdoor systems are covered in this publication.

Any conception of alternative technologies presupposes the integration of information which makes human life healthier and more in harmony with other species and elements of the biosphere. In nature, both small and large cycles exist for the turnover of elements and substances. Human activity often results in the degradation and sometimes the complete breakdown of these cycles, which leads to the accumulation of wastes and pollution of the biosphere. Therefore, the search for alternative forms of energy is urgent for the healthy development of human communities. Humans use non-renewable forms of energy which are located mostly in the lithosphere. This brings about the pollution of air by carbon dioxide and the pollution of water and soil by substances such as iron, aluminum and organic residues. Such ecological degradation modifies the normal succession of biological species and results in a reduction of biodiversity. Contamination of soil and desertification lead to the destruction of the soil eco-communities (a loss of about 20 million hectares per year). The creation of new super-productive forms of crops leads to the leaching of high amounts of nitrogen and phosphorus.

Human communities need to search for new paradigms for living and for ways to integrate human activities into the natural biological cycles of the biosphere. This is the only way to ensure the continuation of our species on this planet. The modeling of such alternative approaches is described in this book.

In general all chapters deal with restorations of biospheric cycles which are uncoupled during the human activity. Therefore this book deals with subjects from molecular biology to human ecology and covers such kind of main subjects like botany, plant physiology, microbiology, biochemistry, soil science, human behavior in the ecosystem. Due to human activity in the discovery of phytohormones and chemical regulators, it was possible to help people achieve the highest of the crop. Ideas of molecular biology in the combination of plant science help now to get the highest productivity of the crops. In the same time, in order to protect soils against destruction, authors proposed unique mechanisms of the application of fabricated soil for landscape rehabilitation.

This book will be of particular interest to biologists, soil scientists, ecologists, agronomists, architects and students of colleges and universities.

**Dr. Narcin Palavan-Unsal**
Istanbul Kultur University
Turkey

# ABOUT THE AUTHORS AND EDITOR

## Biography of V. Kefeli. Dr Sci, PHD

## Valentin I. Kefeli Dr Sci, PhD

I was born in Moscow, Soviet Union, and July 1937. My Mother Alisa (1913-1998) was a chemist. My father Elias (1906-1941) was an architect. My wife Galina is a geographer, Graduated from Moscow Institute of Pedagogical Sciences. My daughter Maria Kefeli - Kalevitsh is Microbiologist, Professor, ViceDean of Robert Morris University, Pittsburgh, PA, USA. In 1958, I graduated from the Agricultural University (Academy) in Moscow as a soil scientist and agro- chemist (Master Degree). From 1961 up to 1996 I worked in Timirjazev institute of Plant Physiology as Assistant (1965- PHD) Researcher, and from 1986, as Professor, Head of the Laboratory of Natural and Synthetic Growth Regulators. Since 1988 I was elected by the staff (600 persons) as President of Institute of Soil Sciences and Photosynthesis, Russian Academy of Sciences, Puschino, and Moscow Region. I participated in Ecological Expeditions in Siberia, Pacific Islands; Sakhalin and Kurily, Central Asia and Polar Regions of Russia.

In 1993, I visited, for the first time, Slippery Rock University and gave two lectures on Modem Russian Economy in European and Siberian Part ( Chair Geography, monitor Dr. Carolyn Prorok). At the same time, I lectured at the Macoskey Center on the problem of plant biomass productivity (Monitor Dr. Larry Patrick, Park and Recreation department). During the Spring Semester, I was a Visiting Professor of SRU, Chair of Biology, discipline- Practical Botany and Biological Seminar. Since 1996 I worked as Associate professor at Department of Biology (Prof. J. Chmielewski) and since 1999 as Manager of Macoskey Center. Since 2000, I worked as Advisor of Slippery Rock Watershed Coalition (Plant- Soil Project in Jennings E.E. Center) Office Ing. Margaret Dunn, Slippery Rock Watershed Coalition, 434 Spring Street Ext, Mars, Pa 16046. At the same time, I work as a volunteer teaching Russian Dialogues (Geography, History and Culture) at SRU, Institute for Learning in Retirement (2007-2009).

# Biogragphy of Winfried E. H. Blum Dipl.-Ing., Dr.rer.nat., Dr.h.c.mult., Prof. of Soil Science

## Winfried E. H. Blum Dipl.-Ing., Dr.rer.nat. Dr.h.c.mult.,

Studies at Universities in Germany and France (1965 M.Sc. Forestry; 1968 PhD Natural Sciences; 1971 Habilitation Soil Science). - 1972-74 Assoc. Prof. for soil science and lecturer for clay mineralogy at the University of Freiburg/Germany. - 1975-79 Visiting Professor and Director of a University Partnership Project at the State University of Paraná in Curitiba/Brazil.

Since 1979 Professor of Soil Science and Director of the Institute of Soil Research, Department of Forest and Soil Sciences at the University of Natural Resources and Applied Life Sciences (BOKU) in Vienna/Austria. Chairman of the Jury of the "Environment and Soil Management Award" of the European Commission together with the European Landowner Organization (ELO), Brussels, Belgium since 2005.

Former activities: Chairman of the Group of Experts on Soil Protection at the Council of Europe, Strasbourg/France (1989-1994). – Secretary-General of the International Union of Soil Sciences (IUSS) (1990-2002). Member of the Scientific Committee of the

European Environment Agency (EEA), Copenhagen/Denmark (1994-2002). Member of the Executive Board, of the Committee on Scientific Planning and Review (CSPR) and Chairman of the Standing Committee "Sciences for Food Security" of the International Council for Science (ICSU), Paris/France (1996-2002). Chairman of the EU Soil Thematic Strategy Working Group "Research" of the European Commission (2002-2004). 2004-2008 Founding President of the European Confederation of Soil Science Societies (ECSSS).

- Co-editor or member of editorial boards of 18 scientific journals. - More than 500 publications in nine languages in the areas of soil chemistry and mineralogy, land use, soil and environmental protection.

- Member or honorary Member of several academies, national and international soil science societies, with numerous distinctions and awards.

- For more information see: http://forschung.boku.ac.at/fis/suche.personen_suchergebn

# Biography of Narçin Palavan-Unsal, Prof. Dr.

## Narçin Palavan-Unsal, Dr Sci, PhD

Professor Dr. Narçin Palavan-Unsal was born in November 3, 1948. She graduated with a bachelor degree in 1971 from Istanbul University, Faculty of Science, Department of Botany and Zoology. In 1971, she served as student assistant in Istanbul University, Faculty of Science, Department of Botany and Zoology. Dr. Palavan-Unsal received her PhD in Plant Physiology from Istanbul University in 1977, and continued her post doctoral research at Yale University on polyamine and plant development relations in 1980 and 1981; and in Munich, at Ludwig Maximillien University on phytochrome pigments.

### Academic Experiences:

1982 - Assistant professor degree

1983 - Associated professor degree

1980/1981 Yale Univertsity (USA) researches on polyamines and plant development

1985/1986 - Munich, Ludwig Maximillien University researches on phytochrom pigments

1988 - Professor degree

2000/2006 - The founder and the head of Halic University Department of Molecular Biology and Genetics

2006- The founder of the Istanbul Kultur University, Department of Molecular Biology and Genetics

Co-editor of 'Advances in Molecular Biology' and 'Journal of Cell and Molecular Biology'

Member of several national and international biology societies, with numerous distinctions and awards.

### RESEARCH INTEREST

- The effect of cyclin dependent kinase inhibitors on cell signaling mechanism in different prostate cell lines.

- The evaluation of polyamines as a biomarker in ovarian cancer.

- The investigation of roscovitin (CYC 202), a Cdk inhibitor in HCT116 colon carcinoma cells in relation with apoptotic signal molecules and polyamine parameters.

- The investigation of the effect of roscovitin (CYC 202) on apoptotic signal molecules in relation with SSAT/polyamine oxidase (PAO) pathway in different breast cancer cell

# ACKNOWLEDGEMENTS

The authors express cordial thanks to all researchers and collaborators who developed the ideas of sun-plant-soil relations. In addition, heartfelt appreciation goes to family members for their passion and support, without which we could not write and publish this book. We also thank to Shari Mastalski and Ajda Coker for technical assistance of the book organization

## CHAPTER 1

# Soil Sustainability

**Abstract:** This chapter deals with general functions of soil in biosphere. The mail conception of soil is the participation with the geological substrate and plants as sources of the organic elements of the soil. Water relations are important components of plant- soil relations. In the chapter are reviewed the elements of landscape sustainability and harmonization of soil uses. Protection of soil during agricultural and industrial use is also discussed.

## SOIL FUNCTIONS

Soil is one of the most important parts of the natural environment and largely non-renewable. World-wide, all economies depend on the goods and services provided by the natural environment. Soils as a natural resource perform a number of key environmental, social and economic functions.

### Ecological Functions

#### *Biomass Production*

Production of biomass ensures food, fodder, renewable energy and raw materials. These well-known functions are the basis of human and animal life [1].

#### *Protection of Humans and the Environment*

Filtering, buffering, and the transformation capacity between the atmosphere, the ground water, and the plant cover, strongly influence the water cycle at the earth's surface, as well as the gas exchange between terrestrial and atmospheric systems. In addition, these functions protect the environment and human beings, against the contamination of ground water and the food chain.

This function becomes increasingly important because of the many solid, liquid, or gaseous, inorganic or organic depositions on which soils react through mechanical filtration, physicochemical absorption and precipitation, or micro-biological and bio-chemical mineralization and metabolisation of organic compounds [2]. These biological reactions may also contribute to global change through the emission of gases from the soil into the atmosphere, because globally, the total pool of organic carbon in soils is about three times higher than the total organic carbon in the above ground biomass and about twice as high as the total organic carbon in the atmosphere.

Under this aspect, soils are a central link in the biotransformation of organic carbon, and they continually play a role in releasing $CO_2$ and other trace gases into the atmosphere. These gases are very important for processes of global change, which in this case, involve large-scale feed-back of many localized, small-scale processes. As long as these filtering, buffering and transformation capacities can be maintained, there is no danger to the ground water or to the food chain. However, these capacities of soils are limited and vary according to the specific soil conditions.

#### *Gene Reservoir*

A biological habitat and gene reserve with a large variety of organisms, soil contains more species in number and quantity than all other above-ground biota together. Therefore, soils are the main basis of biodiversity. Human life is extremely dependent on this biodiversity, because we do not know if we will need new genes for maintaining human life from soils in the near or remote future. Moreover, genes from the soil become increasingly important for many technological, especially biotechnological and bioengineering processes.

### Non-ecological Functions

In addition to the ecological functions, soils have three other functions, more linked to technical, industrial and socio-economic uses.

## *Physical Basis of Human Activities*

Soils are the physical basis for technical, industrial and socio-economic structures and their development, e.g. industrial premises, housing, transport, sports, recreation, dumping of refuse, etc. One of the main problems in this context is the exponential increase of urban and peri-urban areas, including transport facilities between them. This is not only true for Europe, but also for other continents, and especially for countries in development in Africa, Latin America, and Asia.

## *Source of Raw Materials*

Soils are a source of raw materials, e.g. clay, sand, gravel, and minerals in general, as well as a source of energy and water. Raw materials are the basis for technical, industrial, and socio-economic development.

## *Geogenic and Cultural Heritage*

Last but not least, soils are important as a geogenic and cultural heritage, forming an essential part of the landscape in which we live, concealing and protecting paleontological and archaeological treasures of high value for the understanding of our own history, and that of the earth.

In view of the soil as an absolutely limited resource, which cannot be extended or enlarged, the use of these six main functions of soil and land, which is often concomitant in the same area, becomes a key issue of sustainability. Under holistic aspects, soil or land use can be defined as the temporarily or spatially simultaneous use of all these functions, although they are not always complementary in a given area.

## SUSTAINABILITY OF SOIL USE

### Interaction and Competition in the Use of Different Soil Functions

For understanding the role of soil for society and the environment under the aspects of sustainable development, it is necessary to define the interactions and competitions which exist between the uses of soil functions and their uses in space and time. In this context, three different categories of interaction and competition can be distinguished.

### *Soil use for Physical Infrastructure*

Exclusive competition exists between the use of soil for infrastructural development, as a source of raw materials and as a geogenic and cultural heritage on the one hand, and the use of soil for biomass production, filtering, buffering, and transformation activities or as a gene reserve on the other hand. This becomes evident by the sealing of soils through urban and industrial development, e.g. the construction of industrial premises, houses, roads, and sporting facilities, or when soils are used for the dumping of refuse. This process of urbanization and industrialization, excludes all other uses of soil and land. In this context, the exponential increase of urbanization on a world-wide level is one of the main indicators for irreversible soil losses, with unsustainability in soil and land use in the long run. The process of sealing of soils is still very prominent in most of the European countries, and leads daily to severe soil losses.

### *Soil Use for Agriculture and Forestry*

A second category of competition exists through intensive interaction between infrastructural soil/land uses and agriculture and forestry. The intensity of interference, which significantly contributes to the problem of soil contamination and pollution, because all these linear and point sources are loading local soils with contaminants via three different pathways: through atmospheric deposition, on waterways and through terrestrial transport.

There are many possible interactions between infrastructural soil and land use on the one hand and agriculture and forestry on the other hand. This is especially true for densely populated areas in Europe and other densely populated regions of the world. In this context, it also seems necessary to point out that soils are the last but one sink for many inorganic and organic depositions, the last one being the bottom of the oceans. Different forms of loads can be distinguished: inorganic and organic depositions from traffic and transport and from urban and industrial activities. Most of these loads, such as severe acidification, pollution by heavy metals and other elements, pollution by xenobiotic organic compounds, severe salinisation and alcalinisation, as well as deposition of non-soil materials, are more or less irreversible, because souls act as a sink [3]. Irreversibility is defined as the non-reversibility by natural forces or technical remediation measures within 100 years, a time span which corresponds to about four human generations.

Only a few processes of soil degradation, such as superficial compaction or the contamination by biodegradable organics or by small amounts of heavy metals, can be regarded as reversible by technical measures or by natural remediation, e.g. bioturbation and bioaccumulation processes [4].

### *Ecological Soil Use*

A third form of competition also exists among the three ecological soil uses themselves.

Waste and sewage sludge deposition on soils and dredged sediments as well as intensive use of fertilizers and pesticides in addition to the deposition of air pollutants, may have a negative influence on the ground water and the food chain, surpassing the natural capacity of soils for mechanical filtering, chemical buffering, and biochemical transformation. This is specifically true for high input agricultural systems. In this context, it should be remembered that agriculture and forestry do not only produce biomass above the ground but also influence the quality and quantity of the ground water production underneath, because each drop of rain falling on the land has to pass the soil before it becomes ground water.

Such problems are well-known for many parts of the world, where contamination of the ground water through nitrates, pesticides and other chemical compounds from the use of fertilizers, pesticides, and the deposition of sewage sludge and waste compost were analyzed. When the ground water is used as drinking water, the competition between the production of food and fiber above ground and the production of ground water underneath is a competition between the satisfactions of basic human needs.

In many areas of the world, especially in Europe, conventional agricultural production becomes increasingly controlled by quality standards for drinking water. It is easier to transport and sell food and fodder than to do the same with the necessary amount of drinking and household water.

## SUSTAINABLE USE OF SOIL FUNCTIONS

The sustainable use of soils is only possible by a temporal and/or spatial (local or regional) harmonization in the uses of the cited six soil functions, excluding or minimizing irreversible uses, e.g. sealing, excavation, sedimentation, acidification, contamination, or pollution, salinization, and others. This definition includes the dimensions of space and time. The final goal must be to provide multiple functions for the well-being of humans and for the environment.

The necessary harmonization of the uses of the six soil functions is not a scientific but a political issue, which means that all people living in a given area or space should decide which soil functions they will use at a given time and/or at a given space (by a top-down or a bottom-up approach). Scientists only have the possibility to develop scenarios and to explain which causes and impacts may occur when different options are put into function.

## REFERENCES

[1]   Blum WEH, Eswaran, H. Soils for sustaining global food production. Journal of Food Science 2004; 69: 37-42.
[2]   Blum WEH. Agriculture in a sustainable environment-a holistic approach. Int. Agrophysics 1998a; 12: 13-24.
[3]   Blum WEH. Soil degradation caused by industrialization and urbanization. In: Blume HP, Eger H, Fleischhauer E, Hebel A, Reij C, and Steiner KG, Eds. Towards Sustainable Land Use, Advances in Geoecology 31 Icatena Verlag, Reiskirchen. 1998b; pp. 755-766.
[4]   Blum WEH. Soil resilience-the capacity of soil to react on stress. Bollettino della Societa Italiana della Scienza del Suolo 2000; 49: 7-13.

# CHAPTER 2

# Soil Resources and the Environment

**Abstract:** The participation of soil, plants and water in the development of human communities is the main idea of this chapter. The conception of socio-economic metabolism in the environment is discussed in the connection of the interaction of eco-systems and communities. Soil protection in the European Union is described as a model of complex solution of interacting biospheric, factors in socio-eco systems. The factors of the sustainability of soil sphere in the industrial and rural societies is the main idea of this chapter.

## INTRODUCTION

Soil is one of the most important ecosystems, and largely non-renewable. World-wide, all economies depend on the goods and services provided by their natural environment. Soils as a natural resource perform a number of key environmental, social, economic and cultural functions.

Agriculture and forestry depend on soil for the supply of water and nutrients and for root fixation. Soils perform storage, filtering, buffering, and transformation functions, thus playing a central role in the protection of water and the food chain and the exchange of gases with the atmosphere. Moreover, soil is a biological habitat, a gene pool, an element of the landscape, and cultural heritage, as well as the physical basis for human infrastructures, such as houses, industrial premises, transport ways, parking lots, and others. For the installation of these structures, soil provides raw materials.

Growing population and increasing energy consumption, transport, and agricultural activities, linked to pressures, such as global climate variability and warming are adding increasing pressure on the reserve of natural resources in general and especially on the soil environment.

Threats to soil are causing social and economic damage in many regions of the world, amounting to billions of Euros each year, not only reducing the quality of life and the well-being of citizens, but also challenging the social and economic development in those regions at large. Therefore, without the sustainable use of soil, risks and insecurity will increase and economic opportunities will decrease. This underlines the importance of protection and the sustainable use of soils.

In view of the situation in Europe, the European Commission developed a communication to the Council and the European Parliament, entitled: "Towards a Thematic Strategy for Soil Protection," which was ratified by 15 ministers of the environment of the European Union in 2002, European Comission [1].

The purpose of this communication was to build on the political commitment to soil protection in order to achieve a fuller and more systematic approach in the future. With this communication, Europe took a world-wide lead in the commitment to soil protection, because herewith, soil is recognized at the same level as other essential environmental media, such as air and water.

This communication defines the five main functions of soil for human societies and the environment, such as the production of food and other biomass, the capacity for storing, filtering and transformation, soil as a habitat and a gene pool, and soil as a physical and cultural environment for humankind and as a source of raw materials.

Moreover, eight main threats to soils are listed, such as erosion, decline in organic matter, soil contamination (local and diffuse), soil sealing, soil compaction, decline in soil biodiversity, salinization, and floods and landslides. It was also stated that these threats do not apply evenly across Europe, but that there is evidence that degradation processes are getting worse.

Moreover, many European Union (EU) policy areas are of relevance to soil and its protection, especially those related to environment, agriculture, regional development, transport, development and research. It was also stated that knowledge of soil-survey monitoring systems and data networks, and other forms of information are uneven across Europe and that the development of an EU soil protection policy will take time and will require a precautionary approach, based on preventing soil degradation in the future.

**Operational Framework for the Thematic Strategy for Soil Protection**

As a follow-up to the EU-communication, the European Commission, DG Environment, installed five Technical Working Groups (TWG) and an Advisory Form in 2002. These five TWGs addressed all of the eight threats and additionally the issues of monitoring and research. In the following, the primary interest is in the outcome of the TWG Research, which developed new concepts for soil research within the Thematic Strategy for Soil Protection.

## TECHNICAL WORKING GROUP ON RESEARCH

The Research TWG had three mandates: A core specific mandate, dealing with research, and two additional mandates, dealing with sealing and cross-cutting issues.

The results of the TWG Research, with about 65 participants from all over Europe were elaborated during four meetings and a Workshop at the Universität für Bodenkultur in Vienna, from October 28-29, 2004.

During this Workshop, a broad scientific community was invited to discuss the achieved results, and to establish a priority list for research areas in soil protection and the management of Europe's natural soil resources [2].

All the results were based on the DPSIR approach (European Environment Agency) [3], distinguishing between the Driving forces (D), which develop Pressures (P), resulting in a State (S), which by itself creates Impacts (I) and for which Responses (R) are needed. For example, a "D" can be the demand for more space for industrial production, lodging, transport facilities, sports and recreation facilities, the dumping of refuse and others. The "P" deriving from this demand is urbanization in a broad sense, which means the construction of new industrial premises, new houses, and transport ways, such as roads or streets, parking lots and others. The "S" created through this pressure is a sealed soil, which means a considerable loss of agricultural and forest land. The direct "I" is less agricultural (and forest) biomass production, less rainwater infiltration, less biodiversity, and other impacts, with the indirect "I" that farmers have to give up their profession, because there is no land available any more for agricultural or forest production. Moreover, these farmers might move into other areas, causing social and economic problems there. Another indirect "I" of sealing involves the emissions caused by activities on the sealed surfaces, such as air and water pollution, and in some cases also water pollution through concentrated surface water flow under uncontrolled conditions, creating sedimentation and pollution of soils and sediments. The "R" should, whenever possible, be directed at the "D", e.g. towards satisfying the demand for new urban structures by means other than the sealing of new land, e.g. by recycling old, industrially used sites, building on partly polluted or reclaimed old industrial land. This means the "R" would be in the form of incentives, social or economic measures or legal regulation, in order to reduce urban sprawl.

Therefore, the DPSIR framework allows for key questions to be answers in the understanding of complex soil and environmental as well as social, economic and cultural systems, such as:

- What is the "D" behind a problem?
- What is the "Ps" deriving from the Ds?
- What is the "S" which the P creates?
- What are the "Is" that result from the S?

It also allows for Ps to change the Ds, in order to alleviate or reverse the problem, developing solutions through the implementation of operational measures.

Based on this approach, new concepts for research were developed by the TWG Research, to deal with the following specific tasks:

1. Identifying and structuring the existing information;

2. Identifying barriers that prevent the full use of existing results for policies and recommendations on how to improve the transfer of information;

3. Identifying research gaps with indications of the times within which these can be closed (short-, medium-, and long-term).

## PRIORITY RESEARCH AREAS FOR SOIL PROTECTION AND THE MANAGEMENT OF EUROPE'S NATURAL RESOURCES BASED ON DPSIR

This task was accomplished by developing a concept for integrated research in soil resource management and soil protection. In Table **1**, the main research goals, research clusters, and the sciences involved are shown.

The five main research clusters, showing that the analysis of processes, the development, harmonization and standardization of methods for the analysis of the state, the relating of the eight threats to driving forces and pressures, the analysis of the impacts on soil eco-services for other environmental compartments, and finally the development of strategies and operational procedures for the mitigation of the threats, delivering responses, is a necessary consequence of research steps needed to protect the soil resources.

From Table **1** it also becomes clear that not only natural sciences, but also social sciences, economic sciences, historical sciences, philosophical sciences, legal sciences and others have to be involved [2].

**Table 1:** Concept for integrated research in soil resource management and soil protection (from Blum *et al.*, [2].

| Main Research Goal | Research Clusters | Sciences Involved |
|---|---|---|
| 1. To understand the main process in the eco-subsystem soil; induced by threats | Analysis of processes released to the 8 threats to soil and their interdepency; erosion, loss of organic matter, contamination, scaling, compaction, decline in biodiversity, salinisation, floods of landslides | Inter-disciplinary research through cooperation of soil physics, soil chemistry, soil mineralogy and soil biology |
| 2. To know where these processes occur and how they develop with time | Development and harmonization of methods for the analysis of the state (S) of 8 threats to soil and their changes with time = soil monitoring in Europe | Multi-disciplinary research through cooperation of soil sciences with <br> • Geographical sciences <br> • Geo-statistics <br> • Geo-information sciences (eg. GIS) |
| 3. To know the driving forces and pressures behind these processes as related to cultural, social, regional or global developments | Relating the 8 threats to Driving forces (D) and pressures (P) = cross-linking with EU and other policies (agriculture, transport, energy, environment etc.) | Multi-disciplinary research through cooperation of soil sciences with political sciences, social sciences, logistic sciences, historical sciences, philosophical sciences and others |
| 4. To know the impacts on the ecoservices provided by the subsystem soil to other environmental compartments (eco-subsystem) | Analysis of impacts (I) of the 8 threats, relating them to soil eco-services for other environmental compartments: air, water (open and ground water), biomass production, human health, biodiversity | Multi-disciplinary research through cooperation of soil sciences with geological sciences, biological sciences, toxicological sciences, hydrological sciences, physio-geographical sciences, sedimentological sciences and others |
| 5. To have optional tools (technologies) at one's disposal for the mitigation of threats and impacts | Development of opretional procedures for the migration of the threats = Responses (R) | Multi-disciplinary research through cooperation of natural sciences, technical sciences, physical sciences, mathematical sciences and others |

Therefore, at the beginning of the workshop in Vienna, scientists from other areas than soil science were invited to describe and to determine the larger context in which soil protection can be achieved.

## REFERENCES

[1]    Scientific Basis for the Management of European Soil Resources. Research Agenda European Commission 2002; COM, 179 final.

[2]     Blum WEH, Büsing J, Montanarella L. Research needs in support of the European thematic strategy for soil protection. Trends in Analytical Chemistry, 2004b; 23: 680-85.

[3]     European Environment Agency Environment in the European Union at the Turn of the Century. EEA, Copenhagen, Denmark, 1999

# Sustainable Land Management in the Tropics as an Example of Environmental and Socio-Economic Soil Function

**Abstract:** This chapter dedicated to soil sustainability and soil management. One of the most important functions of the soil is production of biomass. This is a link of the soil with solar energy via photosynthesis, Physico-chemical properties of soil components is a basis of biological habitat and gene reserve is the next soil characteristic. Soil is also involved in socio-economic structures. Soil is also geo-genomic and cultural heritage. Tropics were selected as an example of strategy of sustainable soil management in tropics

## INTRODUCTION

In recent years, "sustainable land management" has become a key word in the discussion about agricultural land use, especially in food production in the tropics. Discussions at international congresses, seminars and workshops on this topic reveal considerable conceptual differences in the basic approach. Many scientists define "land management" as "agricultural land management," without taking into consideration that at least five other main uses of soil and land interact competitively with agriculture in space and time.

Therefore, in the following, an attempt will be made to define sustainable land management through a comprehensive approach, distinguishing between "sustainable land management" and "sustainable agricultural land management," identifying direct and indirect indicators as well as problems of sustainable agricultural land management in the tropics on a strategic, technical, and operational level, including general approaches and future tasks [1-9].

## THE SIX MAIN FUNCTIONS OF SOIL AND LAND

In a comprehensive approach, six main uses of soil and land can be distinguished: Three of them ecological, and three others mainly related to human activities, such as technical, industrial and socio-economic ones. The three ecological uses are:

1.  Production of biomass, ensuring the supply of food, fodder, renewable energy and raw materials; a function which is of basic importance for human and animal life, and is regarded with quite different priorities in tropical countries compared to countries of the northern hemisphere.

2.  Filtering, buffering and transformation capacity of soil between atmosphere, groundwater and plant cover. Soils act as a protective medium, preventing the uptake of harmful substances by plant roots and their translocation into the ground-water, producing at the same time, gases through biochemical transformation processes. These filtering, buffering and transformation activities which depicts three main processes:

    *   Mechanical filtration, e.g. in the pore space,

    *   Physico-chemical buffering through adsorption and precipitation on the surfaces of inorganic and organic soil components,

    *   Transformation by microbiological and biochemical processes, especially the decomposition and alteration of organic compounds through mineralization and metabolisation.

Observations of the global distribution of organic carbon reveal that its total pool in soils is three times higher than that in the above-ground biomass, and twice as high as the total organic carbon in the atmosphere [10]. Therefore, soils are a central link in the biotransformation of organic carbon, and continually play a role in releasing $CO_2$ and other trace gases into the atmosphere. These gases are very important for processes of global change, which in this case involve large-scale feedback of many localized small-scale processes. Therefore, the management of beneficial soil biota in tropical agroecosystems is very important.

3.  Moreover, soil acts as a biological habitat and a gene reserve, because a greater variety of organisms lives in the soil than in the above-ground biomass. Therefore, soil use is directly linked to biodiversity, which is also an important factor for human life, considering for example, the fact that the antibiotic penicillin was developed from the ubiquitous *Penicillium* fungus, present in the soil. We do not know if we will need new genes for maintaining human life in the future. In addition, genes from the soil are increasingly used for biotechnology and biogenetic engineering. Soils also have three technical, industrial and socio-economic functions.

4.  Soils are the spatial basis for technical, industrial and socio-economic structures and their development, e.g. for industry, housing, transport, sports, recreation, dumping of refuse, and others.

5.  Soils serve as a source of geogenic energy, raw materials (e.g. clay, sand, gravel etc.), and water.

6.  Finally, soils are a geogenic and cultural heritage, which form part of the landscape, and conceal and preserve paleontological and archaeological treasures, which help us understand our history and that of the earth.

Six main soil functions are below:

1.  Agricultural and forest production,

2.  Infrastructure

3.  Filtreing, buffering transformation

4.  Gene reserve and protection

5.  Geogenic and cultural heritage forming landscapes

6.  Source of raw materials

These six functions indicating that land use should be defined as the temporally and spatially simultaneous use of all these soil functions, which are, of course, not always complementary in a given area. But where specific conditions on a regional or local scale exist, this very broad and comprehensive definition can be reduced or adapted to those conditions. Moreover, severe competition and interaction exist between these six main uses of soil and land.

## COMPETITION BETWEEN THE SIX MAIN SOIL FUNCTIONS AND USES

For developing a definition for sustainable land management it is necessary to define all the interactions and competitions which exist between these six functions and their uses. Three main types of competition and interaction can be distinguished.

Exclusive competition exists between the use of soil for infrastructure, as a source of raw materials and as a source of geogenic and cultural heritage on the one hand, and biomass production, filtering, buffering, and transformation activity and as a gene reserve on the other hand. This is quite obvious as agricultural production, filtering, buffering and transformation activity or natural gene reserve exists under roads or factories.

At the same time, there are intensive interactions between infrastructural land uses and their development, and agricultural and forest land use. Dense road and settlement infrastructure contributes significantly to the problems of soil contamination.

These linear and point sources load local soils with contaminants on three different pathways: through atmospheric deposition, on waterways (e.g. groundwater or irrigation water), and through terrestrial transport. This is especially through for densely populated areas, which are growing exponentially in developing countries (Tables **1** and **2** )

Agglomerations in tropical areas are growing faster than those in industrialized contries. Therefore, urban and periurban development and its connecting transport infrastructure are very alarming problems for sustainable land management. More than two-thirds of the total world population is born in those agglomerations, with an increasing tendency.

**Table 1:** Growth of urban populations between 1970 and 1990 (United Nations, 1991/92).

| Area | 1970 (%) | 1990 (%) |
|---|---|---|
| Europe | 67 | 73 |
| South America | 60 | 76 |
| North America | 58 | 71 |
| Africa | 23 | 34 |
| Asia | 24 | 29 |
| World | 37 | 43 |

**Table 2:** Increase of the world's 35 largest cities between 1970 and 1990 (population in millions).

| No | Country | City | 1970 | 1990 | % Growth |
|---|---|---|---|---|---|
| 1 | Japan | Tokyo | 14.91 | 20.52 | 37.6 |
| 2 | Mexico | M.City | 9.21 | 19.37 | 112.4 |
| 3 | Brazil | S.Paulo | 8.22 | 18.42 | 124.1 |
| 4 | USA | N.York | 16.29 | 15.65 | -.3.9 |
| 5 | China | Shanghai | 11.41 | 12.55 | 10.0 |
| 6 | India | Calcutta | 7.12 | 11.83 | 66.2. |
| 7 | Argentina | B. Aires | 8.55 | 11.58 | 35.4 |
| 8 | Korea | Seoul | 5.42 | 11.33 | 109.0 |
| 9 | India | G.Bombay | 5.98 | 11.13 | 86.1 |
| 10 | Brazil | Rio de J. | 7.17 | 11.12 | 55.1 |
| 11 | UK | London | 10.59 | 10.57 | -.0.2 |
| 12 | Japan | Osaka | 7.61 | 10.49 | 37.8 |
| 13 | USA | Los Ang. | 8.43 | 10.47 | 24.2 |
| 14 | China | Peking | 8.29 | 9.74 | 17.5 |
| 15 | Indonesia | Jakarta | 4.48 | 9.42 | 110.3 |
| 16 | Russia | Moscow | 7.07 | 9.39 | 32.8 |
| 17 | Iran | Teheran | 3.29 | 9.21 | 179.9 |
| 18 | Egypt | Cairo | 5.69 | 9.08 | 59.6 |
| 19 | France | Paris | 8.34 | 8.75 | 4.9 |
| 20 | India | N.Delhi | 3.64 | 8.62 | 136.8 |
| 21 | Philippines | Manila | 3.6 | 8.4 | 133.3 |
| 22 | China | Tiajin | 6.87 | 8.38 | 22.0 |
| 23 | Italy | Milan | 5.52 | 7.9 | 43.1 |
| 24 | Pakistan | Karachi | 3.14 | 7.67 | 144.3 |
| 25 | Nigeria | Lagos | 1.44 | 7.6 | 427.8 |
| 26 | Thailand | Bangkok | 3.27 | 7.16 | 119.0 |
| 27 | USA | Chicago | 6.76 | 6.89 | 1.9 |
| 28 | Peru | Lima | 2.92 | 6.5 | 122.6 |
| 29 | Bangladesh | Dhaka | 1.54 | 6.4 | 315.6 |
| 30 | India | Madras | 3.12 | 5.69 | 82.4 |
| 31 | Colombia | Bogota | 2.37 | 5.59 | 135.9 |
| 32 | Hong Kong | Hong Kong | 3.53 | 5.44 | 54.1 |
| 33 | Russia | St. Petersburg | 3.96 | 5.39 | 36.1 |
| 34 | Iraq | Baghdad | 2.1 | 5.35 | 154.8 |
| 35 | Spain | Madrid | 3.37 | 5.06 | 50.1 |

Finally, intensive competition also exists among the three ecological soil uses. Agricultural soils are contaminated by waste and sewage sludge deposition, as well as by intensive use of fertilizers and plant protection products, in addition to the overall contamination through infrastructural land uses.

This should be taken into account when implementing high input agricultural systems under tropical conditions, because in many cases, the natural capacity of soils for mechanical filtering, chemical buffering, and biochemical transformation can be exceeded. In this context it should be remembered that farmers are producing biomass (food, fodder and renewable energy) on their land, but, at the same time, groundwater underneath, because each drop of rain falling on their land has to pass the soil before it becomes groundwater. Therefore, farmers influence not only the food chain, but also the quantity and the quality of groundwater produced, through their agricultural practices, especially the use of agrochemicals. Intensive agricultural land management influences the gene reserve as well as the biodiversity.

To summarize, it can be stated that in many parts of the world, including the tropics, agricultural land use is influenced by the growth of urban and periurban development, as well as by industrial growth. Therefore, a more comprehensive definition of "sustainable land management" is needed.

### Definition of Sustainable Land Management

Sustainable land management can be defined as spatial and/or temporal harmonization of all the six main uses of land, minimizing irreversible ones, e.g. sealing, excavation, sedimentation, contamination or pollution, salinization, alcalinization and others. This definition includes the dimensions of space and time, and aims at maintaining a maximum of uses in a given area, thus leaving options of land management for future generations.

This harmonization has to be attempted in a local or regional approach, depending on the specific uses in a given area. The achievement of this harmonization is not possible without political intervention. Scientists can contribute by developing scenarios to facilitate sound political decisions.

### Definition of Sustainable Agricultural Land Management

According to the comprehensive definition of sustainable land management given above, agricultural land management is only one of several possible land uses, and therefore depends on all other uses in a given area or region. Thus, sustainable agricultural land use is only possible, when all the other types of land use are sustainable as well. In this definition sustainability is not only determined by ecological, but also by socio-economic and cultural factors. In many cases socio-economic factors are of higher importance than ecological prerogatives. These socio-economic conditions are normally determined on a global or on a regional level, such as the GATT negotiations, regularly held at various places in the world. The socio-economic factors are governed by market conditions, agreements on tariffs and trade and others, especially the cost of energy, raw materials and labor. In many cases, these are the predominant factors for high input agricultural systems in the tropics

In contrast, ecological factors are defined on a local scale, taking into account specific physical, chemical, biological, soil and other conditions for agricultural production.

### INDICATORS FOR SUSTAINABLE AGRICULTURAL LAND MANAGEMENT

Indicators for sustainable agricultural land management can be of an ecological, as well as socio-economic and cultural nature.

### Ecological Indicators

Ecological indicators are mostly of a physical, chemical, and biological nature, and should be based on a farm or local level (except for the exchange of gases, which is difficult to control on such a scale). Such indicators could be: soil characteristics, including products which pass through the soil into the groundwater, or products, which are produced above the soil, such as biomass and gases. The solid, liquid and gaseous phases of soil must be controlled to ensure sustainability including groundwater characteristics, because these are also influenced by agricultural land management.

## Socio-economic and Cultural Indicators

These indicators are market conditions, agreements on tariffs and trade, costs of labor, energy and raw materials, as well as cultural conditions, such as education and training systems and others, which are normally not based on a farm or local level but are defined in terms of geopolitical interests on regional or even global levels.

## Indicator Use

Using these two main groups of indicators, it seems possible to establish a systematic approach for indicators within a given area or region.

According to the prevailing agricultural production system and the intensity of use of fossil energy and raw materials, two extremes of systems can be defined in agricultural production: low and high input systems.

When checking ecological as well as socio-economic and cultural indicators on the basis of this approach, it becomes clear that ecological or biological indicators, such as human health, species diversity, or others, have very different indicative values, when applied to high-input and low-input systems.

Human health may be a very good indicator for sustainability in low input systems, because the conditions of health and nutrition indicate, how well basic needs, such as food and water for the local population, are satisfied at the farm or local level.

In high-input systems, such as those in Europe or North America, human health.

Can not be used as an indicator, because the satisfaction of basic needs does not depend entirely on local farming conditions, but rather on the existing trade, and on other socio-economic parameters.

Species diversity, on the other hand, is a good indicator for both, high-input and low-input systems.

The definition of indicators can only be based on their prevailing functions within a given agricultural production system. This means that "soil health" or "soil quality" cannot be indicators, because they are too broad. Therefore, soil characteristics need to be analyzed on the basis of their role in the controlling processes within these systems. The question then arises, as to which soil functions dominate in a given situation. For example, soil can be analyzed in relation to its functions as a biological habitat, as a porous system, or as a biomass production system through the availability of water, air and nutrients, and can also be analyzed for specific transport processes between atmosphere and groundwater. Therefore, in general, all soil indicators have to be based on their specific functions in controlling processes.

The use of physical, chemical and biological threshold values is only possible for local or site conditions. Threshold values applied to regional or global levels are ecologically meaningless. For example, with regard to groundwater contamination, a gravelly or sandy soil cannot have the same threshold values for nitrogen input as a clay soil, because both soil systems are ecologically and functionally very different. Threshold values of broad regional scale can have intensive political impacts in the sense of education on regulative measures, but these can be of questionable local significance.

When using indicators for the analysis or assessment of sustainability, it seems necessary to carefully check the system for which they should be analyzed. Considering the costs, time and infrastructural requirements which would be needed for the analysis of soil parameters, and considering that many soil characteristics cannot be monitored so exactly as to indicate short-time changes, it may be necessary to consider soil fluxes as potential indicators. The assessment or analysis of fluxes or loads, using an input/output model, under certain circumstances, can provide more useful indicators than using direct, costly measurements of specific soil parameters.

Several existing models in soil ecology allow us to predict impacts on soil systems through adverse fluxes or loads, and this can help us to define sensitive indicators when needed.

Data bases for the definition of sustainable land management are already available in many areas of the world. Therefore, new research priorities in this direction should only be defined after careful evaluation of already existing data. Moreover, more use of existing soil information in the form of soil maps or soil data based on geographical information systems is necessary for the analysis and development of sustainable agricultural land management systems.

System analyses on local, regional or global levels have very different goals, and are not always compatible, because the reliability of data differs enormously in scale and time. The output of such analyses has to be carefully checked.

Long-term requirements for sustainable agricultural land management are conditioned in the bounds of nature and can be summarized as the essential governing principles of the biosphere, which should be respected when defining the framework for different sustainable land management systems.

## PROBLEM IDENTIFICATION FOR SUSTAINABLE AGRICULTURAL LAND MANAGEMENT IN THE TROPICS

The identification of problems for sustainable agricultural land management in the tropics can be achieved on three different levels: on a strategic, tactical and operational level.

### Strategic Level

Based on the holistic approach developed above, several important problems in relation to good production in the tropics should be considered.

The growth of urban and periurban areas, with regard to sustainable soil and water management. Urbanization in developing countries, especially in the tropics, has reached a very dynamic stage and a magnitude which is without any historical precedent. Two-thirds of the world population growth occurs exclusively in cities. If this development continues, it will cause severe impacts on farmland areas, with important consequences for agricultural production.

The question of a further extension of agricultural land (cropping areas and pasture) and its influence on soil protection, based on our actual knowledge, it seems evident that there are no more soil reserves for agricultural use available in the tropics, because any further extension of agricultural and pasture land will lead to the destruction of natural terrestrial ecosystems and soils [11].

In many areas of the tropics, competition between food and groundwater production exists. Therefore, primary goals have to be established in order to maintain equilibrium in the satisfaction of both basic needs.

Besides the definition of these and further problems, which have to be considered in order to achieve sustainable agricultural land management in the tropics, the procedures under which this should occur have to be redefined as well.

There is a need for defining new approaches, based on the essential governing principles of the biosphere. Such basic long-term requirements for sustainable agricultural land management can be defined as follows:

1.  Maintenance of the greatest possible variety of species in and upon the soil (biodiversity), and protection of water resources. The old and traditional land use techniques, which were developed in the tropics and are adapted to their specific ecologic conditions, should be given priority, and the many techniques that are imported from other ecological regions should be adapted accordingly.

2.  Solar orientation and the use of energy from renewable resources with minimized use of fossil energy and non-renewable raw materials is the need of the hour, especially in view of the difficulties for providing those inputs in tropical regions due to socio-economic and other constraints.

3.  Optimal use of energy and raw materials through the combination of different land use systems (e.g. agroforestry), and closed-cycle production, with a reduction of material intensity and an increase of resource productivity, is another important target which should be defined as a basic procedure for sustainable agricultural land use in the tropics.

4.   Decentralization in combination with ecological land management systems, and maintenance of ecologically stable landscapes is another principle which seems to be very important in order to maintain soil fertility and landscape productivity, avoiding their destruction.

5.   Agreements on specific market conditions, tariffs and terms of trade for agricultural products, avoiding eco-dumping. This principle seems to be very important and is far beyond science, reflecting problems in the socio-economy of many tropical countries.

Based on these procedures, the targets of entropy reduction, minimization of entropy production and increase of ecological stability can be reached. Therefore, it seems necessary to increase our efforts for the development of new concepts of sustainable agricultural land management of the tropics in the near future [12].

## Tactical Level

Whereas the strategic level is mainly based on a global or continental approach, the tactical level deals with regional and local problems. Within this level, ecological, socio-economic and technical constraints have to be reconsidered.

Ecological constraints for sustainable land management in the tropics are the limited surfaces with fertile soils and their constant reduction through soil deterioration, especially through erosion by water and wind, salinization and alcalinisation. This means that the maintenance and improvement of soil fertility is one of the most important tactical targets in tropical land management. Other constraints are: topography, climatic conditions, availability of water resources, a lack of natural vegetation for the protection of agricultural land due to deforestation, and other forms of severe mismanagement in tropical landscapes.

Socio-economic and technical constraints are the lack of administrative and technical infrastructures on a country level, and within specific regions. This also indicates that administrative and technical support for farmers is strongly underdeveloped.

The same is true for legal instruments, protecting landscapes and soils, and promoting sustainable land management (personal experience), through maintenance of certain price levels for agricultural products.

Economic instruments e.g. incentives for the stimulation of sustainable production systems are rarely available. Moreover, research is insufficient in many countries, or its results are not available at the farmer's level. The same is true for education and information, which are the basic tools for promoting sustainable agricultural land management in the tropics.

## Operative Level

Operative procedures have to focus on local or site conditions. Under such conditions, different constraints can be identified in relation to sustainable agricultural land management in the tropics:

•   Availability of information at the farmer's level, e.g. through extension work.

•   With the available labor on the decline, in many regions there is a need for mechanization to counterbalance the loss of manpower, especially in those rural areas where migration to urban and periurban centers is common.

•   Moreover, the availability of funds and products, e.g. fertilizers and agrochemicals for providing favorable growth conditions and protecting crops, and agricultural products, is lacking in many areas.

Such constraints at an operative level cannot be solved at a scientific level, as they are mostly based on socio-economic or on specific legal conditions.

## APPROACHES TO SUSTAINABLE AGRICULTURAL LAND MANAGEMENT IN THE TROPICS AND FUTURE TASKS

With regard to the actual situation, four steps should be taken in order to meet the challenge of sustainable land management in the tropics. These steps are:

- Definition and identification of Problems: Within this target, ecological as well as socio- economic problems should be identified and distinguished clearly.

- Monitoring of changes: It seems absolutely necessary that the different countries concerned with sustainable agricultural land management carefully monitor changes in urban as well as in rural areas, including migration processes, losses of soils and the protection of natural vegetation covers.

- Controlling by corrective measures:

- Based on points discussed earlier, corrective measures are introduced by way of socio-economic measures, as well as by improving agricultural land management techniques on a scientific basis.

- Rehabilitation and remediation: In many areas rehabilitation and remediation work has to be introduced in order to save agricultural land for sustainable management, and in order to counterbalance the daily losses of soils and landscapes.

## Important Future Research Targets for Sustainable Agricultural Land Management in the Tropics

Those can be formulated on the basis of the above-defined requirements and considering the actual state of knowledge:

Definition of criteria for sustainable production systems for local or regional ecological units on the basis of the governing principles of biological systems;

Development of criteria for the definition of ecologically stable landscapes;

Assessment of ecologically tolerable bearing capacities of landscapes, regarding adverse atmospheric, aquatic and terrestrial constraints;

Definition of the crucial links between agricultural and other forms of land management, especially technical and industrial production systems.

## REFERENCES

[1]  Blum WEH. Sustainable land management with regard to socioeconomic and environmental soil functions. A Holistic Approach. In: Wood RC and Dumanski J. Eds. Proceedings of the International Workshop on Sustainable Land Management for the 21st Century. Volume 2: Plenary Papers. 1994a; pp. 115-24. The Organizing Committee. International Workshop on Sustainable Land Management. Agricultural Institute of Canada, Ottawa.

[2]  Blum WEH. Sustainable land use for food production in the tropics and Subtropics. A Holistic Approach. J. für Entwicklungspolitik X. 1994b; 3: 301-314.

[3]  Blum WEH. Sustainable land use for sustainable food production in Africa, Proceedings of a Seminar organized by COSTED-IBN in Accra/Ghana, 5-6. April 1994, COSTED-IBN, Madras/India, 1994c; pp. 92-107.

[4]  Blum WEH. Sustainable land use and environment. Proceedings of the Diamond Jubilee Symposium "Management of Land and Water Resources for Sustainable Agriculture and Environment" of the Indian Society of Soil Science, New Delhi, India; 1994d; pp. 21-30.

[5]  Blum WEH. A concept of sustainability, based on soil and soil functions. Proceedings of the Soil Fertility Research Institute Bratislava/Slovac Republic, 1995; 19/1: 3-13. Proceedings of the Conference to the 35th Anniversary of the Institute "From Soil Survey to Sustainable Farming", October 3-5, 1995 in Stara Lesna, High Tatras.

[6]  Blum WEH. Sustainable Land Use, a Comprehensive approach to control land degradation. Programme, abstracts and excursions. 1996; pp. 4-5, Internal Conference on Land Degradation, June 10-14, 1996, Adana/Turkey, Cukurova University Press, Adana.

[7]  Blum WEH. Sustainable land use: A Holistic Approach. In: Sehgal J., Blum WEH and KS. Gajbhiye Eds. Red and Lateritic Soils. Vol. 1: Managing Red and Lateritic Soils for Sustainable Agriculture.1998a; pp. 10-21, Oxford abd IBH Publ. Co., New Delhi, Calcutta.

[8]  Blum WEH. Sustainability and Land Use. In: D'Souza GE and TG. Gebremedhin Eds. Sustainability in Agricultural and Rural Development. 1998b; pp. 171-191, Ashgate, Aldershop UK, Brookfield USA, Singapore, Sydney.

[9]    Sehgal J, Blum WEH, Gajbhiye KS. Red and Lateritic Soils Vol. 1: Managing Red and Lateritic Soils for Sustainable Agriculture. Oxford and IBH Publ. Co. New Delhi. United Nations Environmental Data Report (3$^{rd}$ Ed.), 1998

[10]   Eswaran H, Van den Berg E, Reich P. Organic carbon in soils of the world. J Soil Sci Soc Am 1993; 7: 192-94.

[11]   Oldeman LR, Hakkeling RTA, Sombroek WG. World map of the status of human-Induced soil degradation: An explanatory note. ISRIC-UNEP Report, Nairobi, Kenya. 1991.

[12]   Blum WEH, Santelises AA. A concept of sustainability and resilience based on soil functions: In: Greenland DJ and Szabolcs I. Eds. The Role of the International Society of Soil Science in Promoting Sustainable Land Use. Soil Resilience and Sustainable Land Use. 1994. pp. 535-542, Cab International, Wallingford.

# CHAPTER 4

## Fabricated Soil for Landscape Rehabilitation

**Abstract:** Land protection could be carried on by the fabricated soil as a tool for landscape restoration. The composition of this soil is based on the structure of natural profile and on the analytics of natural spoil components and their properties. The main idea of the construction of FS is carried on proper ratio of carbon – nitrogen components as well as supply of water retaining aluminum-silicate materials

## INTRODUCTION

Fabricated soil (FS) is a mixture of decaying substrates rich in aluminosilicate, carbon, nitrogen, phosphorus and potassium sources. This substrate usually is used for landscape rehabilitation. After the exposure of FS soil into the natural habitat for the duration of one year, the microbial activity certainly changed remarkably.

FS are developed from a mixture of materials, which encourage plant development. FS in this study were developed for use in the reclamation of drastically disturbed lands. One of the main components of native soil as well as FS is the aluminosilicate matrix provided by clays (illite, smectite, kaolinite, etc.) formed as weathering products of such minerals as orthoclase and other feldspars, and micas, such as muscovite (high potassium content), biotite and others. These minerals are necessary contributors of calcium, magnesium, sodium, and iron [2]. The size consistency (soil texture) of the mineral fraction in native soil varies from clay-size to coarse sand-size. The carbon- and nitrogen-rich organic matter contains the monomers and polymers, the main constituents of the humus complex. Sources of cellulose are dry leaves and sawdust, which also provide lignin. These constituents are humus precursors. Humus is more or less a stable fraction of soil organic matter. It sorbs the mineral nutritive elements of nitrogen, potassium and phosphorus, which are important for plant growth and development. Natural soils are commonly described through soil profiles.

FS is an important tool for landscape restoration. Soil is considered as the protective screen of life [1]. Soil is a necessary intermediate substrate in the regulation of the biosphere activity. The loss of soil resources increases up to 10-15 million hectares in a year. Therefore, rehabilitation of the soil cover is a global problem that could be solved by cooperation of such disciplines as mineralogy, soil science, biology, ecology, agrochemistry, and biochemistry. Developing public-private partnership efforts to utilize (recycle) local waste is promising with both economic and environmental benefits.

The natural decay of readily-available parts helps to enrich the soil with nutrients at least for certain duration of time necessary to maintain a healthy balance in the soil.

The proposed recipes of FS are based on the conception of carbon-nitrogen balance in the soil as well as on the transformation of carbon products such as glucose, phenolics and plant polymers–cellulose and lignin in the humus polymer that is tightly connected to the aluminosilicate matrix of the soil micelle. The role of microorganisms in the composting process is important. They combine their activity with plants in the transformation of plant organic substances. Before coming to the conception of the FS we investigated the structure of the native soil profile typical for Western Pennsylvania, Gresham Soil.

## SOIL PROFILE

Soil is a dynamic natural body composed of mineral and organic materials and living forms in which plants grow; therefore, the fertility is usually tested by the growth activity of plants. Soil is organized in nature in layers or soil horizons. These horizons are approximately parallel to the ground surface and are different in properties and characteristics from adjacent layers. A vertical section of the soil through all its horizons is the soil profile. Horizons are associations with common characteristics like hydrology, mineral and organic components, specific weight, color, thickness, structure, etc. The minimal element of soil structure is the soil micelle, a fine particle consisting of aluminosilicate crystals (clay minerals) with adsorbed ions and humus molecules.

Alfisols are productive soils, having medium to high base saturation, with good crop yields. They have a generally favorable texture and are located in zones with sufficient rainfall. Typically, the surface layer (8 inches) is a dark grayish brown silt loam. The subsoil is mottled to a depth of 75 inches. The upper 13 inches are brown to yellow brown friable silt loam, and the next 18 inches are firm. Permeability is moderately slow. The lower 36 inches are firm and brittle silt loam and gravelly loam. The seasonal water table is 6-18 inches below the surface and runoff is slow or medium. Up to 40 inches in depth, the acidity is very strong. Clay content = 15-32%; Moist bulk density = 1.3-1.9 %; Organic matter = 2-4% (Butler, County Soil Survey, 1989) (Table **1**).

## FABRICATED SOIL COMPONENTS

The process has been started to create an artificial soil-like mixture using four readily available components in ample quantity. These are: topsoil (Gresham-type), from the excavation of the wetland treatment system and experiments; the upper layers of the pond sediments, from the excavations of the algae and plant ponds; "humanure" compost aged one year, as a product of habitation; and 5-7 year-old, partially-decomposed wood chips, from adjacent pathways (Table **2**).

**Table 1:** Characteristic soil horizons.

| Horizon | Depth | Description |
|---|---|---|
| A | 0-28 cm (11 in.) | Gray, sandy, non-plastic with many roots and organic residues |
| E | 28-52 cm (20 in.) | Subsoil, ochre, loamy, some roots, non-stick, non-plastic |
| E/B | 52-62 cm (24 in.) | Transitional horizon, ochres with gray fragments, no roots, more plastic, compact |
| B | 62-93 cm (37 in.) | Illuvial horizon, agrillic, mostly gray, plastic, stick |
| B/C | 93 cm | Compact, cemented horizon with rock fragments |

\* Soil to the B/C horizon will be available for use in the fabricated soil.
\* Site-specific soil horizons

**Table 2:** Horizons to be used in fabricated soil.

| Horizon | Description | % Organic Matter | Specific Density |
|---|---|---|---|
| A<br>0-28 cm | Brownish gray,<br>Non-sticky<br>SOM-rich<br>Not dense<br>**Crumbly** | 0.65 | 1.23 g/cm |
| E<br>28-52 cm | Yellow gray,<br>Red-orange flecks<br>Sticky<br>SOM-poor<br>Dense<br>**Not crumbly** | 0.04 | 1.25 g/cm |

The components are presently available in the following estimated quantities (given in cubic feet).

1. Topsoil-72

2. Upper layer of pond sediments-264

3. Humanure compost-152

4. Wood chip mulch-234

More topsoil will become available after the excavation of the wetland treatment basins. The lower layers of pond sediments may be used to supplement the material from the upper zone, if needed, as there appears to be an excess of material needed to line the enhanced basin. The amount of material from the lower zone is estimated at 784 cubic feet.

Preliminary testing was conducted on all components of the mixture to identity the presence of chemical and physiological properties, bacteriological flora, and general fertility. Research is on-going in order to refine the recipe to provide the most productive mixture for outdoor plants (established in the ponds and irrigation field), indoor plants (for cloning and enhancement of the indoor wetland and sanitary plants), and platelets for the greenhouse and seed beds (Tables **3** and **4**).

**Table 3:** Biological activity (length of shoots, cm).

| Plant | Horizon | 1 | 2 | 3 | 4 | 5 | Av |
|-------|---------|---|---|---|---|---|-----|
| Clover | A | 6.5 | 6.5 | 6.4 | 6.3 | 6.3 | 6.4 |
|       | E | 5.5 | 5.5 | 5.3 | 5.3 | 5.3 | 5.4 |
| Lettuce | A | 7.1 | 1.0 | 6.8 | 6.8 | 6.7 | 6.7 |
|       | E | 6.0 | 5.9 | 5.8 | 5.7 | 5.6 | 5.7 |

**Table 4:** Soil nutrients.

| Horizon | N | K | P |
|---------|---|---|---|
| A | Low to medium | High | High |
| E | Trace | Medium | Medium |

## CHARACTERISTICS OF FS COMPONENTS

### Topsoil

0-28 cm, Gresham silt loam, non-plastic, with organic residues, a good medium for germination and seedling growth, trace nitrates, organic matter 5.6%, trace phosphates, medium to high potassium, and high microbial activity.

### Pond Sediments

Silt clay loam, pH 7.0-7.5, trace nitrates, medium to high phosphates, high to very high potassium, organic matter 5.6 %, high biological activity, good substrate for seed germination and growth, usable as a component for rooting substrate, good water retention.

### Compost

Rich in microflora, low level of *E. coli,* high fertility, high in carbon, only partially decomposed and potential source of macronutrients N, P, K (composted cow manure could be substituted for composted human manure).

### Wood Chip Mulch

A black humus-rich substrate that assists in soil aggregation and increased porosity. A soil conditioner used for growth of plant regenerates.

The creation and utilization of the Macoskey agrisol complements the many human activities around the center. It is the culminating product that brings many byproducts into the closed biospheric recycling: excrement, saw dust, and excavated material.

## BIO-MINESOIL

Another FS is being developed through a partnership with a non-profit organization (Stream Restoration Inc.), a limestone quarry operator (Quality Aggregates Inc.), and a watershed group (Slippery Rock Watershed Coalition). This soil is being fabricated to aid in promoting the use of local waste materials including wastes from abandoned and active mine operations. This FS is based on the combination of two sources of carbon substances: readily-composted leaves (cellulose) and slowly-composted sawdust (lignin). Experiments were conducted during the spring and summer of 2001. The final composition is as follows (by volume):

1 part - light gray limestone fines, source of nutrition

1part - gray pond sediments, water retention, organic matter, potassium

1 part - abandoned coal refuse

1 part - cow manure, nitrogen, phosphorus source

1part - dry leaves (source of carbon, humus substrate)

1 part - sawdust

1 part - potting soil (slowly composted carbon source, humus substrate)

1 part – sand.

Specific weight of this type of FS is 1 $g/cm^3$.

This is a balanced soil composition with the carbon sources: sawdust (long splitting part) and leaves (easy splitting part). It stimulates early crop growth as tested on lettuce, mustard, wheat and clover. Lawn herbs developed similarly in the "BioMinesoil" as in the control-potting soil. This soil is much cheaper than commercial soil composition as the waste materials are free, except for short distance hauling and handling costs.

The experiment with leaves was conducted in the cold frames of the Macoskey Center, Slippery Rock University. Each cold frame plot was divided into two parts: with and without maple leaves. Varieties of mint and lavender were planted in each plot. All experimental data were recorded in September 2001 (Tables **5** and **6**).

Previously, we discussed the presence of heterotrophic microflora in different components of the FS. Top-soil and compost were very rich in the heterotrophic microflora compared to other components of the FS, like pond sediments, leaves, or saw dust.

**Table 5:** Effect of maple leaves on the herbs mass. Control: fabricated soil without maple leaves.

| Plant | % to Control |
|---|---|
| Lavender | 119 |
| Apple mint | 147 |
| Chocolate mint | 182 |
| Lemon mint | 165 |

**Table 6:** Effect of maple leaves on length of plant stem Control: Fabricated soil without maple leaves.

| Plant | % to Control |
|---|---|
| Lavender | 140 |
| Apple mint | 150 |
| Chocolate mint | 88 |
| Lemon mint | 117 |

After one year exposure of the FS into the natural environment in Western PA, it was determined that the amount of mold was declining but the amount of bacterial heterotrophic microflora (BHM) was on the rise. The highest amount of BHM, more than 5,900,000 colonies per plate, was observed for the red willow (*Salix rubrum*) plot. Poplar plot showed a plate count of 5,700,000. These were compared with the original sample of FS that contained 170,000 colonies per plate. All calculations were done per gram of soil. In the original FS sample from 2002, the number of the mold colonies was 360,000 per gram, but a year later, on the poplar plot, the mold count dropped drastically to 23,000. As for the willow plot, it was 270,000 mold colonies per gram of soil. Poplar and willow are active producers of phenolics - derivatives of salicilate. Their root excretions might act as natural antiseptics, thus bringing numbers in mold colonies down.

The balance of carbon and nitrogen was not remarkably changed during the one year period and was still on average C: N - 10:1. Transformation of FS in nature is also based on the presence of different species of wild plants grown on these plots. Among the wild plants, clovers were dominant. They had well-developed root nodules (bacteroids).

Therefore the population of BHM depends not only on the dead substrates of FS but also on the nitrogen-fixing bacteria in legumes or pioneer plants in this case. Natural nitrogen fixation is an important part of soil health.

Dry leaves from different trees that were buried in the FS exuded phenolics within one month. More stable ones were coumarins and less stable ones were others such as flavonoids and antocyanins. Thus BHM may play an important role in the transformation of phenolic inhibitors (allelopathogens) by maintaining healthy soil-plant relations and emphasizing the importance of FS use in landscape rehabilitation

In general, the process of wood formation in plants is tightly connected with the further humus formation in the soil. These biological processes make FS a cover/top soil for abandoned mine sites as a step in landscape remediation. The components of soil health are numerous and abundant.

## REFERENCES

[1]    Kovda Y. Problems of soil protections and planet biosphere. Kefeli V. Ed. Puschino, ONTI, 1989; p. 156.

[2]    Brady NC. Nature and properties of soils. McMillan Publ. Comp. NewYork, 1984.

## CHAPTER 5

# Microbial Activity in Fabricated Soils for Landscape Rehabilitation

**Abstract:** This chapter deals with biological activity of fabricated soil (F). Effect of FS on plant growth, microbial cenosis, as well as interaction of bacteria and fungi during FS development. Thus, bacteria and fungi are considered here as factors of plant residue transformation (necropolis). Very important are components of microbial cenosis which split carbon polymers such as cellulose, starch and lignin.

## INTRODUCTION

Fabricated soil (FS) is a mixture of organic and inorganic components to create a substrate rich in aluminosilicates, carbon, nitrogen, phosphorus and potassium. A year after incorporation into coal mine soil, dramatic increases in bacterial heterotrophic microflora (BHM) and equally dramatic decreases in mold colonies were observed. Decreases in mold colonies were particularly noted in plots containing poplar and willows whose root excretions appear to act as natural antiseptics. BHM may, in turn, play an important role in the transformation of phenolic inhibitors (allelopathogens) thus maintaining healthy soil-plant relationships and emphasizing the role of fabricated soils in landscape rehabilitation.

FS biological activity was determined since 2002 up to 2004 and compared with the other substrates (Table **1**).

**Table 1:** Biological activity of FS (Fabricated Soil) and other substrates. Percentage of seed germination (G %) and seedling length in mm (L mm) were determined for seedlings of 4 crops after 1 week exposure: temperature 22°C, Light 15 feet/chandelles, and luminescent lamps. Standard error 5%.

| Substrates | Wheat | | Clover | | Mustard | | Lettuce | |
|---|---|---|---|---|---|---|---|---|
| | G % | L mm | G % | L mm | G % | L mm | G % | L mm |
| Potting Soil | 78 | 46 | 100 | 45 | 77 | 49 | 40 | 33 |
| FS start 2002 | 55 | 15 | 89 | 10 | 75 | 18 | 39 | 13 |
| Mining substrate | 43 | 10 | 59 | 10 | 27 | 5 | 28 | 10 |
| Sand | 58 | 20 | 67 | 22 | 47 | 11 | 40 | 13 |
| FS-2004 Poplar plots | 64 | 46 | 96 | 34 | 76 | 37 | 63 | 24 |
| FS-2004 willow plots | 65 | 45 | 96 | 32 | 76 | 39 | 82 | 24 |

FS in two years of being exposed in nature became more fertile and closer in their properties to the fertility of potting soil.

## MICROBIAL ACTIVITY OF FS

Just as fungi are the most important soil microbiota in retaining nutrients in forest soil, bacteria are of comparable importance in agricultural soils and grasslands. Bacteria retain nutrients first in their biomass, and second, in their metabolic by-products. Numbers of total bacteria generally remain the same regardless of soil type or vegetation [1].

The study that was conducted in 2004 identified certain numbers of bacterial colonies found in soil samples [2, 3]. Poplar (*Populus balsamifera*) plot of FS showed maximum numbers of CFU and a variety of bacterial species. The data of 2005 evidenced the increase of the bacterial amounts in all samples in comparison with the 2004 year (Table **2**).

During the one-year exposure of FS into the natural environment, the activity of bacterial communities changed the growth of trees, whereas the green mass increased remarkably. These trees developed deep root systems, and the level of nutrients such as potassium, nitrogen and phosphorus was also on the rise. Thus, FS became an enrichment substrate for plant growth and propagation, and for microbial activity, as the interrelations between all living components of the soil constitute successful strategies in fighting against soil degradation and erosion. Fabricated or so-called "artificial soils" are high in fertility and can be used in the areas where erosion and loss of topsoil is evident. Examples of possible applications could include areas of former coal mining and stripping activities, or areas of steel production, which unfortunately determined heavy, environmental impacts in Western Pennsylvania [1]. Natural composting or decay of old maple leaves and mushrooms also induced the growth of fungi populations. The process of leaves composting was reduced after one year of FS processing in nature. This process continues in 2005 (Table **3**).

**Table 2:** Number of bacterial colonies found in soil samples.

| Soil type | Total Bacterial Count, CFU x $10^6$ / number of bacterial cells | |
|---|---|---|
| | 2004 | 2005 |
| Mining soil | 0.0022/8 | 0.063 |
| Top soil | 0.43/8 | 711 |
| Fabricated soil | 27/9 | 711 |
| Fabricated soil 2002 poplar | 66/10 | 68.4 |
| Fabricated soil 2002 willow | 18/8 | 486 |
| Fabricated soil 2003 poplar | 114/14 | 495 |
| Fabricated soil 2002 willow | 30/8 | 882 |

*All data are statistically significant. 95% confidence intervals exist for all points. A = 0.05 [4].

**Table 3:** Fungal count in soil samples.

| Soil type | Fungal count, CFU/gram* | |
|---|---|---|
| | 2004 | 2005 ($10^6$) |
| Mining soil | 36,000 | 0.18 |
| Top soil | 34.000 | 0.81 |
| Fabricated soil | 290,000 | 6.1 |
| FS poplar plot (*Populus nigra*) 2002 | 510,000 | 9 |
| FS poplar 2003 | 120,000 | 6 |
| FS willow (*Salix discolor*, Muhl) 2002 | 220,000 | 9.8 |
| FS Willow 2003 | 41,000 | 2.7 |

**Table 4:** Microbial components in fabricated and mining soils, 2007.

| Bulk samples | Microbes | Seed Plot | Trees Plot |
|---|---|---|---|
| | | Percentages (%) | |
| Fabricated Soil | Total bacterial count | 76.5 x $10^6$ CFU/gram | 82.8 x $10^6$ CFU/gram |
| | *Sphingomonas paucimobilis* | 0 | 50 |
| | *Bacillus* spp*. 3 | 50 | 50 |
| | *Bacillus* spp*. 7 | 50 | 50 |
| | *Arthrobacter* spp. | 50 | 50 |
| | *Bacillus* spp*. 5 | 50 | 50 |
| | *Bacillus* spp*. 4 | 50 | 50 |
| | *Bacillus* spp 6*. | 0 | 50 |
| | *Chryseobacterium indologenes* | 0 | 50 |
| | *Rhizobium radiobacter* | 0 | 50 |
| | | | 2007 |
| **Bulk samples** | **Microbes** | | **Percentages (%)** |
| Mining soil | Total bacterial count | | 3.600 CFU/gram |
| | Unidentified Actinomycetes | | 50 |
| | *Bacillus* spp*. 3 | | 25 |
| | *Bacillus* spp*. 7 | | 25 |

*More than one species were recovered from sample.

By monitoring soil organisms' dynamics, we can detect detrimental ecosystem changes and possibly prevent further degradation. The response of each group of soil organisms, i.e. soil saprophytic bacteria, symbiotic bacteria, saprophytic fungi, mycorrhizal fungi, protozoa, and nematodes, can be used to indicate effects of contaminants on soil health [1, 4]  (Table **4**).

FS is certainly an excellent substrate for fungi development. Maple leaves, compost and topsoil are components of FS formed by the dynamic system of fungal cenosis. It is important to mention that the interaction of fungi and bacteria during the natural decay of FS brings an array of nutrients to plants, insects and other representatives of bioflora to enrich the environment in general, and soils in particular [4] (Table **5**).

**Table 5:** Percentage of bacterial presence in tested soil samples of fabricated soil.

| Bacterial Species | % |
|---|---|
| *Bacillus* spp 2 | 15 |
| *Bacillus* spp 3 | < 1 |
| *Bacillus* spp 6 | 4 |
| *Bacillus* spp 1 | 11 |
| *Bacillus* spp 4 | 11 |
| *Actinomycete* | 36 |
| Coryneforum gram-positive *Bacillus* | 7 |
| Gram-positive coccus | 4 |

In FS the most abundant fungi found was *Penicillium* spp. up to 58%, followed by *Aspergillus* 17%.

## REFERENCES

[1]     Kalevitch MV and Kefeli VI. Study of bacterial activity in fabricated soils. Int J Environment and Pollution 2007; 29: 412-423.

[2]     Kalevitch MV, Kefeli VI, Borsari B, Davis J and Bolous G. Chemical signaling during organism's growth and development. Journal of Cell and Molecular Biology 2004a.

[3]     Kalevitch MV, Kefeli VI, Dunn M, Johnson B, Taylor W, Borsari B. Bacterial activity in fabricated soils. Presentation at 104th American Society of Microbiology General Meeting, New Orleans. 2004 b

[4]     Kalevitch M, Kefeli V. Plant Biodiversity in the fabricated soil experiment. Journal of Sustainable Agriculture 2006; 29: 3.

# CHAPTER 6

# Microbial Activity of Fabricated Soil and Plant Biodiversity

**Abstract:** Higher plants form their cenosis later than the process of the development of the microbial eco-system. Grassy plants are the sources of organic matter for the Fabricated Soil (FS) as well as sources of nitrogen substances which were formed in the plant nodules of roots of Fabaceae, legumes. About 20% of nitrogen could come to FS thanks to roots nodules activity (Alfa Alfa, clovers and other). The composition of plant patterns on FS could be changed during the season.

## INTRODUCTION

Fabricated soil (FS) is a natural mixture of decaying substrates rich in alumni-silicate, carbon, nitrogen, phosphorus, and potassium sources. This substrate usually is used for landscape rehabilitation and is an excellent source and example of environmental remediation. We studied the use and application of FS in Western Pennsylvanian soils, areas that were previously degraded by acid mining drainage. We are still trying to determine if fabricated soils are a long-term or short-term solution to the problem. This soil was tested on its fertility and then applied to mining substrate. After one year, the first weed plants on the soil appeared in May 2003. The composition of the weeds on the fabricated plots was determined. After the exposure of FS into the natural habitat for the duration of three years, we also evaluated bacterial and fungal activity in the soil as this is an important indicator of soil health.

FS is a composition of inert and nutritive substrates for landscape rehabilitation. This soil was tested on its fertility and then applied to the mining substrate in Jennings Environmental Education Center, Slippery Rock, PA, in May 2002. In one year, the first weed plants on the soil appeared in May 2003. The composition of weeds on the fabricated plots was determined. New forms and the amount of each form of weeds were investigated on the poplar and willow plantations.

Plant biodiversity depends on different factors: climate conditions, water supply, allelopathic relations between species, and soil properties (physio-chemical and biological). We observed the development of weed populations on the FS, which we applied on the inert mining substrate. The weed pattern was dependent on the time of the season and on the process of spreading the seeds over the FS.

An increasing interest has emerged with respect to the importance of microbial diversity in soil habitats. The extent of the diversity of microorganisms in soil is seen to be critical to the maintenance of soil health and functions. The two main drivers of soil microbial community structure are plant type and soil type. They are thought to exert their function in a complex manner. It is believed that in some situations the soil type is the key factor in determining soil microbial diversity, in others, it is the plant type. Both are related to the complexity of the microbial interactions in organisms and plants [1].

Certain bacterial strains, for example, are particularly important in nitrogen cycling [2-4]. Free-living bacteria fix atmospheric nitrogen, adding it to the soil nitrogen pool. Other nitrogen-fixing bacteria form association with the roots of leguminous plants such as lupine, clover, alfalfa, and others (National Science and Technology Center). Needless to say, legumes, clovers in particular, were used purposefully to grow on the FS plots in our experiment.

It was stated by several researchers that little information is available concerning the occurrence of natural transformation of bacteria in soil. This is because few bacteria are known to possess the genes required to develop competence and because the tested bacteria are unable to reach this physiological state *in situ* [5].

Highly productive agricultural soils tend to have ratios near one for bacterial: fungal biomass. Soil is one of the most commonly studied fungal habitats. They are entirely dependent upon the kind of material falling into the soil. Thus a fungus specializing in oak leaves will probably not colonize a pine needle. This factor alone will account for many differences between the mold populations of two soils. In addition, fungi may be sensitive to moisture levels, pH, and competition from other organisms, and many other influences [6, 7] (where mold are found).

Interrelations of fungal mycelium with other soil biota are of paramount importance in soil ecology. The variables studied included a number of soil properties, bacteria, protozoan flagellates, ciliates and amoebae, microbial and plant feeding nematodes, various microarthropods, and two fungal biomarkers [8].

The extent to which soil resource availability is linked to patterns of microbial activity and plant productivity within ecosystems has important consequences for our understanding of how ecosystems are structured and for the management of systems for agricultural production. One of the examples that correlate with our research was a study done in southwest Michigan, United States, on a site that had been cultivated and planted with row crops for decades. Soil samples were analyzed for physical properties (texture, bulk density), chemical properties (moisture, pH total C, total N, inorganic N), and biological attributes (microbial biomass, microbial population size, respiration potential, and nitrification and N-mineralization potentials). Plant analyses included biomass and C and N contents. Soil resource variability across this long-cultivated site was remarkably high, as was variability in microbial activity and primary productivity. Overall, results suggest a remarkable degree of spatial variability for a pedogenically homogeneous site that has been plowed and cropped mostly as a single field for over 100 years. Such variability is likely to be generic to most ecosystems and should be carefully evaluated when making inferences about ecological relationships in these systems and when considering alternative sampling and management strategies.

## THE EXPERIMENT: MATERIALS AND METHODS

The biological activity of FS was determined by placing soil samples in Petri dishes. Each variant contains three Petri dishes with soil. Potting soil (commercial version) was used for the standard. Ten grams of soil was placed in one Petri dish and then it was watered by 6 ml water. Seeds of four crops were placed in four separate sectors of the Petri dish. We determined the germination and seedling growth of wheat var. "Genova", mustard, "Southern Giant Curled", lettuce, "Black Seeded Simpson", and red clover, "Red Crimson" [9, 7]. FS was applied on the mining substrate area in May 2003 in forms of plots 2 meters by 2 meters. On each plot 25 young (one-year) cuttings were planted of poplar (*Populus nigra*), black willow (*Salix nigra* Marsh), and pussy willow (*Salix discolor* Muhl). Weeds were collected from these plots since 2003 every month during the season. They were dried for herbarium and identified by illustrated Flora for Northern United States, NY, 1913. In 2004 weeds were also counted and the dominating species were determined.

## RESULTS AND DISCUSSION

Transformation of FS in nature is also based on the presence of different species of wild plants grown on these plots. Among the wild plants, clovers were predominant and they had well-developed root nodules (bacteroids). After the one-year exposure of the FS to the natural environment in Western Pennsylvania, it was determined that the amount of mold was declining but the amount of bacterial heterotrophic microflora (BMH) was on the rise. Therefore, the population of BHM might also depend on the nitrogen-fixing bacteria in legumes or pioneer plants in this case, as natural nitrogen fixation is an important part of soil health.

The poplar plot showed a plate count of $5.7 \times 10^6$ CFU/g soil. All plots were compared with the original sample of FS that contained $1.7 \times 10^5$ CFU/g soil. In the original FS sample from 2002, the number of the mold colonies was $3.6 \times 10^5$ CRU/g soil, but a year later on the poplar plot, the mold count dropped drastically to $2.3 \times 10^4$ CRU/g soil [10]. As for the willow plot, it dropped to only $2.7 \times 10^5$ CFU/g soil. Poplar and willow are active producers of phenolics-derivatives of salicilate. Their root excretions might act as natural antiseptics thus bringing numbers in mold colonies down. Thus, FS is a good substrate for the development of microbial cenosis. The study that was conducted in 2004 identified certain numbers of bacterial colonies found in soil samples [11, 12]. Poplar (*Populus balsamifera*) plot of FS showed maximum numbers of CFU and a variety of bacterial species [13] (Table **1**).

Most of the samples had different *Bacillus* species. Nine types were identified. *Bacillus* represents a genus of Gram-positive bacterium, which is ubiquitous in nature (soil, water, and airborne dust). This bacterium has a unique characteristic in that it has the ability to produce endospores when environmental conditions are stressful. Although most species of *Bacillus* are harmless saprophytes, two species are considered medically significant: *B. anthracis* and *B. cereus* [14-19,2,3] (Table **2**).

**Table 1:** Number of bacterial colonies found in soil samples.

| Soil Type | *Total bacterial count CFU x $10^6$ | Number of Bacterial Species |
|---|---|---|
| Mining soil | 0.0022 | 8 |
| Top soil | 0.43 | 8 |
| Fabricated soil | 27 | 9 |
| Fabricated soil 2002 poplar | 66 | 10 |
| Fabricated soil 2002 willow | 18 | 8 |
| Fabricated soil 2003 poplar | 114 | 14 |
| Fabricated soil 2003 willow | 30 | 8 |

* 95 % of confidence intervals exist for all points, a= 0.05.

**Table 2:** *Bacillus spp.* types.

| Type | Description |
|---|---|
| *Bacillus* spp. Type 1 | Translucent, spready colony, irregular edge, gram-positive rods with endospores, motile |
| *Bacillus* spp. Type 2 | Dry white colony. Gram-positive rods with endospores. Motile (*Bacillus cereus*) |
| *Bacillus* spp. Type 3 | Large, white spready colony, swirling over the surface of the agar. Gram-positive rods with endospores. Motile (*Bacillus mycoides*) |
| *Bacillus* spp. Type 4 | Round translucent colony with smooth edges. Gram-positive rods with endospores. Motile. |
| *Bacillus* spp. Type 5 | Spreading, yellow moist colony with irregular edges. Gram-positive rods with endospores. Motile. |
| *Bacillus* spp. Type 6 | Wrinkled, small off-white colony with irregular edges. Gram-positive rods with endospores. Motile. |
| *Bacillus* spp. Type 7 | Spreading clear, large colony that splatters across the agar surface. Gram-positive rods with endospores. Motile. |
| *Bacillus* spp. Type 8 | Moist white colony with irregular edges. Gram-positive rods with endospores. Motile. |
| *Bacillus* spp. Type 9 | Small, moist, round, cream colored colony with irregular edges. Gram-positive rods with endospores. Motile. |

*Bacillus cereus* was the common representative in different types of tested soils. Its presence varied from 29% in 15% in freshly formed FS. Future two-year exposure of FS to the natural environment showed the increase in the percentage of this bacterium up to 24% and 43%, respectively, on poplar and willow plots. This indicated that FS offer an excellent solution for improving soil fertility and landscape restoration as the result of interrelations between chemical by-products of decay, plant, and microbial activity [13].

*Actinomycetes* are also another important group of soil organisms that promote soil health. They were absent in mining soil but present in topsoil by 5%. Original FS had 36%. *Actinomycetes* are great decomposers so, logically, after two years of natural decay for organic matter in fabricated soils, the percentage dropped to 6 and 3 respectively at the poplar and willow plots [13].

Bacteria are becoming increasingly important in bioremediation, meaning that we can use them to help in the reclamation of contaminated lands [20, 21]. Bacteria are capable of filtering and degrading a large variety of human-made pollutants in the soil and ground water so that they are no longer toxic. The list of materials they can detoxify, just to name a few, includes herbicides, heavy metals, and petroleum products [2, 3].

By monitoring soil organism dynamics, we can detect detrimental ecosystem changes and possibly prevent further degradation [18]. The response of each group of soil organisms, that is, soil saprophytic bacteria, symbiotic bacteria, saprophytic fungi, mycorrhizal fungi, protozoa, and nematodes, can be used to indicate the effects of contaminants on soil health [13].

During the one-year exposure of FS into the natural environment, the activity of bacterial communities changed the growth of trees. Whereas the green mass increased remarkably, these trees developed deep root systems, and the level of nutrients such as potassium, nitrogen, and phosphorus was also on the rise. Thus, FS became an enrichment substrate for plant growth and propagation, and for microbial activity as the interrelations between all living components of the soil constitutes successful strategies in fighting against soil degradation and erosion [13].

The number of fungal species in different soil types differed depending on several factors, including the year of the research (Table **3**). FS had the most variety in fungal species compared even with topsoil, and certainly mining soil had the least number of fungal species, only two. The most prevalent fungi in soil samples were *Trichoderma* and *Penicillium*. The full array included *Fusarium, Mucor, Aspergillus,* and more.

**Table 3:** The number of fungal species in different soil types.

| Soil Type | Number of Fungal Species* |
|---|---|
| Mining soil | 2 |
| Top soil | 5 |
| Fabricated soil | 6 |
| FS poplar plot 2002 | 5 |
| FS poplar 2003 | 7 |
| FS willow 2002 | 6 |
| FS willow 2003 | 4 |

*95% of confidence intervals exist for all points, a= 0.05.

FS on the poplar plot 2002 had mostly *Penicillium spp.* 39%, *Acremonium* 20%, *Trichoderma and Fusarium* 14% and 12% respectively. FS poplar plot 2003 had *Aspergillus* 33%, *Trichoderma and Fusarium* both 17%, and *Penicillium spp.* and *Mucor* both 8% [13].

*Mucor* mold, a filamentous fungus found in soil, plants, decaying fruits and vegetable, was also present. As well as being ubiquitous in nature, it is a common laboratory contaminant.

Willow 2002 Plot had *Trichoderma* 41%, *Acremonium and Fusarium* spp. 14% both, followed by *Penicillium* spp. 7% and 5%, respectively. Most of the fungi found in the soil samples are part of a normal soil flora [13].

FS is certainly an excellent substrate for fungi development. Maple leaves, compost, and top soil are components of FS formed by the dynamic system of fungal cenosis. It is important to mention that the interaction of fungi and bacteria during the natural decay of FS brings an array of nutrients to plants, insects, and other representatives of bioflora to enrich the environment in general, and soils in particular [13] (Table **4**).

**Table 4:** Fungal count in soil samples.

| Soil Type | Fungal Count, CFU (gram*) |
|---|---|
| Mining soil | 36,000 |
| Top soil | 34,000 |
| Fabricated soil | 290,000 |
| FS poplar plot (*Populus nigra*) 2002 | 510,000 |
| FS poplar 2003 | 120,000 |
| FS willow (*Salix discolor*, Muhl) 2002 | 220,000 |
| FS willow 2003 | 41,000 |

*95% of confidence intervals exist for all points,
a= 0.05.

FS biological activity was determined from determined during the time 2002-2004 and compared with the other substrates (Table **5**).

**Table 5:** Biological activity of FS (fabricated soil)-2002.

| Substrates | Wheat | | Clover | | Mustard | | Lettuce | |
|---|---|---|---|---|---|---|---|---|
| | G % | L mm | G % | L mm | G % | L mm | G % | L mm |
| Potting soil | 78 | 46 | 100 | 45 | 77 | 40 | 40 | 33 |
| Fabricated soil start-2002 | 55 | 15 | 89 | 10 | 75 | 18 | 39 | 13 |
| Mining substrate | 43 | 10 | 59 | 10 | 27 | 8 | 28 | 10 |
| Sand | 58 | 20 | 67 | 22 | 47 | 11 | 40 | 13 |
| FS-2004 poplar plots | 64 | 46 | 96 | 34 | 76 | 37 | 63 | 24 |
| FS-2004 willow plots | 65 | 45 | 96 | 32 | 75 | 39 | 82 | 24 |

FS in the two-year exposure in nature became more fertile and more closely resembled the fertility of potting soil. The results showed that mining substrate was the least fertile and showed the least amount of growth. Clover was the most abundant on all of the substrates. Only lettuce showed more intensive growth on the 2004 poplar and willow plots than on regular potting soil.

Now we approach the development of vascular plants-autotrophs on the same substrate.

## WEEDS ON THE FABRICATED SOIL DURING THE SEASONS

The first weed plants on the plots with FS appeared in May 2003. Among them were dominated wild geranium (*Geranium maculatum* L.), birds foot trefoil (*Lotus corniculatus* L.), red clover (*Trifolium pretense* L.), yellow water cress (*Radicula palustris* L.), corn speedwell (*Veronica arvensis* L.), and ground ivy (*Glechoma hederacea* L.). All these weeds inhabited the plantations of fast growing poplar and willow, which were developed on the FS. In June 2003, one month later on the same plantation, the additional weeds of lady's thumb, heartweed (*Polygonum pericarpa* L.), ox-eye daisy (*Chrysanthemum leucanthemum* L.), and Kentucky blue grass appeared. In July 2003, we also observed ringed panic grass (*Panicum annulum* Ashe), slender golden rod (*Solidago erecta* Pursh), and common plantain (*Plantago major* L.) (Table **6**).

**Table 6:** Weeds on the FS during the different seasons-2003.

| Month of the year | Weeds (grassy plants) |
|---|---|
| May | 1.Dominated wild geranium-*Geranium maculatum* L.<br>2.Birdsfoot trefoil-*Lotus corniculatus* L.<br>3.Red clover-*Trifolium pretense* L.<br>4.Yellow watercress-*Radicula palustris* L.<br>5. Corn speedwell-*Veronica arvensis* L.<br>6. Ground ivy-*Glechoma hederacea* L. |
| June | 1.All weed found in May<br>2.Ladys's thumb, heartweed-*Polygonum persicaria* L.<br>3.Oxeye daisy-*Chrysanthemum leucanthemum* L.<br>4.Kentucky Bluegrass-*Poa pratensis* L. |
| July | 1.All weed found in May and June<br>2.Ringed panic grass-*Panicum annulum* Ashe<br>3.Slender goldenrod-*Solidago erecta* Pursh<br>4.Common plantain-*Plantago major* L. |

During the year 2004, we observed new species of weeds in May: Woody yarrow (*Achillea lanulosa* Nutt); in June, pink deptford (*Dianthus armeria* L.), diffuse cinquefoil (*Potentilla millegrana* Engelm), wood meadow grass (*Poa crocata* Mich), and large flowered spear grass (*Poa eminens*); and in July, new species of ribgrass (plantain-*Plantago lanceolata* L.) (Table **7**).

## COMPOSITION OF WEED PATTERN ON FABRICATED SOIL PLOTS

In 2004 the amount of weeds on the poplar and willow plots increased remarkably. The composition of weeds was checked before and after weeding.

**Table 7:** Weeds on the FS during the different seasons, 2004.

| Month of year | Weeds (grassy plants) |
|---|---|
| May | 1.Woody yarrow-*Achillea lanulosa* Nutt |
| June | 1.Pink depthford-*Dianthus armeria* L.<br>2.Diffuse cinquefoil-*Potentilla millegrana* Engelm.<br>3.Wood meadow grass-*Poa crocata* Mich.<br>4.Largeflower speargrass-*Poa eminens* |
| July | 1.New species of plantain, ribgrass-*Plantago lanceolata* L. |

Wild geranium was dominant among the April weeds on the poplar and willow plots (Table **8**).

**Table 8:** Weeds on the plots with the FS in April 2004 (before weeding)-in percent to the total amount of the weeds on the plot.

| Plots of FS | Weeds (grassy plants) | | | | | | |
|---|---|---|---|---|---|---|---|
| | Daisy | Wood meadow grass | Trefoil | Clower | | Wild geranium | Pink deptford |
| | | | | Red | White | | |
| Poplar 2002 | | 28 | | 18 | 14 | 18 | 16 |
| Poplar 2003 | | 3 | | | | 80 | |
| Willow 2002 | | 16 | 1 | 22 | 18 | 2 | 1 |
| Black willow 2003 | | | | 3 | | 61 | |

In June 2004, we observed on the different plots, mostly daisy, wild geranium. When the plots were weeded in April, the wild geranium was still dominating on the seeded plots, but the pink deptford also was wide spread (Table**9**).

**Table 9:** Weeds on the plots with FS in June 2004 (no weeding) in percent of the total amount of weeds on the plot.

| Plots of FS | Weeds (grassy plants) | | | | | |
|---|---|---|---|---|---|---|
| | Daisy | Wood meadow grass | Wild geranium | Pink depthford | Red clower | Yarrow |
| Poplar 2002 Plot 1 | 25 | | 25 | 50 | | |
| Poplar 2002 Plot 2 | 6 | 17 | 24 | 30 | | |
| Poplar 2002 Plot 3 | 20 | 7 | 20 | 46 | | |
| Black willow | 25 | | 16 | 16 | 8 | |
| Pussy willow | 36 | 31 | | | 9 | 18 |

These data showed that the biodiversity of weeds on the plots of FS do not depend much on the woody plants which were grown on the plots. The season had more influence on the weed composition. Plants with seeds which could be spread by the wind were more actively developed on the plots.

The data for weeds might be correlated with the microbial composition of the FS. During the processing of FS in nature, the composition of microflora as well as the composition of plant weeds changed remarkably [11, 12, 22].

Also, it is important to mention the *Fabaceae* species: clovers and trefoils which possess the ability of nitrogen fixation by visible nodules. These plants could participate in soil nitrogen stabilization [11, 12].

Boulos [24] showed that plants are able to interact by chemical signaling. This effect might also be observed on FS, during the investigation of weed composition on the plots.

Data from this study suggest that plant biodiversity on FS can be used for landscape rehabilitation in areas that were previously used for mining and stripping. New growth of various plants was seen over the course of a few years and new ones could continue to be seen in the future.

# REFERENCES

[1]    Garbeva P, Van Veen JA, Van Elsas JD. Microbial diversity in soil: Selection of microbial populations by plant and soil type and implications for disease suppressiveness. Annual Review of Phytopathology. 2004; 42: 243-270.

[2]    Ingham ER, Trorymow JA, Ames RN, Hunt HW, Morley CR, Moore JE, Coleman DC Trophic interactions and nitrogen cycling in a semiarid grassland soil. Part 1. Seasonal dynamics of the soil food web. J Appl Ecol. 1986a; 23: 608-615.

[3]    Ingham ER, Trol'ymow IA, Ames RN, Hunt HW, Morley ER, Moore LE, Coleman DE. Trophic interactions and nitrogen cycling in a semiarid grassland soil. Part II. System responses to removal of different groups of soil microbes or fauna. Journal of Applied Ecology, 1986b; 23:615-630.

[4]    National Science and Technology Center. Bureau of Land Management. Soil Biological Communities. Web site: http://www.b!m.gov/llstdsoillbacterial.

[5]    Demaneche S, Kay E, Gourbiere F, Simonet P. Natural transformation of *Pseudomonas jluoroscens* and *Agrobacterium tumefaciens* in soil. Applied and Environmental Microbiology, 2001; 67: 2617-2621.

[6]    Kefeli V, Kalevitch M Natural growth inhibitors and phytohormones The Netherlands: Kluwer Publishers. 2002; pp. 340.

[7]    Kefeli V, Kalevitch M. Phenolic cycle in plants and environment. Journal of Cell and Molecular Biology. 2003; 2: 13-18.

[8]    Krivtsov V, Grillilhs BS, Salmond R, Liddell K, Garside A, Bezginova T, Thompson A., Staines, Watling R, Palfreyman JW. Some aspects of interrelation between fungi and other biota in forest soil. Mycological Research. 2004; 1: 933-946.

[9]    Kefeli V. Nawro! growth inhibitors and phytohormones. Holland. 1978.

[10]   Where Moulds are Found. n.d. Web site: http://www.botany.uforonfo.calResearch Luhs/M

[11]   Kalevitch M, Kefeli V, Borsari B. Bacterial activity in fabricated soils. Presentation at I04th American Society for Microbiology General Meeting, New Orleans. 2004a.

[12]   Kalevitch M, Dunn M, Borsari B, Taylor W, Johnson D, Kefeli V. Microbial activity in fabricated soils. Abstracts of 104 General meeting, American Society for Microbiology, New Orleans, Louisiana, N-OJO, Soil Microbiology, Agronomic Sec. 2004b, pp. 399.

[13]   Kalevitch M, Kefeli V. Study of bacterial activity in fabricated soils. Int J Environment and Pollution, 2007; 24: 412-423.

[14]   Bongers T. The Maturity Index: An ecological measure of environmental disturbance based on nematode species composition. Oecologia, 1990; 83: 14-19,

[15]   Coleman DC, Odum EP, Crossley DA Jr. Soil biology, soil ecology and global change. Biol Fert Soils 1992; 14: 104-111.

[16]   Coleman DC. Through a ped darkly: An ecological assessment of root-soil-microbial-fauna interactions In: Fitter AH, Atkinson D, Read DJ, and Usher MB, Eds. Ecological Interactions in Soil. Blackwell Scientific Publications, Cambridge, 1985; pp. 1-21.

[17]   Dindal D. Soil Biology Guide. John Wiley and Sons. 1990; pp. 1349.

[18]   Ingham ER. Web site: htTp://www.rain.org/-sals/inghall1.htl1ll

[19]   University of Texas-Houston Medical School n.d. Web site: http://medic. med. uth. Ime. edll/palh/OOOO /437.hlm

[20]   Bogan BW, Sullivan WR, *et al.* "Humic coverage index" as determining factor governing strain-specific hydrocarbon availability to contaminant-degrading bacteria in soils. Environmental Science & Technology Easton, 2003; 37: 5168-5174.

[21]   Griftiths BS, Ritz K, Ebblewhite N, Dobson G. Soil microbial community structure: Effects of substrate loading rates. Soil Biol Biochem. 1999; 31: 145-153.

[22]   Kefeli V. Practical Botany. Slippery Rock University, USA. 1995.

[23]   Bolous G, Borsari B, Kefeli V. Botanical herbicide in action. Journal of scholarly Endeavor, 2004a; 4:7.

[24]   Bolous G, Borsari B, Kefeli V. The effect of sumac (*Rhus* spp.) leaves and roots water soluble phenolics on the rooting of willow and bean cuttings. Slippery Rock Watershed Coalition, 2004b:7.

# CHAPTER 7

# Nitrogen Fixation as a Source of Nitrogen in the Soil

**Abstract:** Thus, air components are the sources of biomass formation on the FS. Carbon dioxide is the starting gas for the plant photosynthesis and nitrogen of air participates in the process of nitrogen fixation. The biosynthesis of nitrogen containing products- amino acids, proteins and alkaloids is a result of the activity of nitrogen- fixing organisms. The activation of the process of nitrogen fixation is based on gene modification of plant – bacterial symbiosis.

## INTRODUCTION

It is a well-known fact that growth and development as well as productivity itself depends on the availability of mineral nutrients. Nitrogen is one of the major elements, which is required in large amounts as an essential component of proteins, nucleic acids and other cellular constituents.

78% of the nitrogen is present in the atmosphere in the inert form. However, its availability for plant life is based on the presence of "fixed" forms such as ammonium ($NH_4$) or nitrate ions. Microbes play detrimental role in nitrogen availability and support of life on the planet. Both bacteria and fungi play important roles in nitrogen fixation. Nitrogen fixation uses bacteria that are either free-living or form symbiotic associations with plants or other organisms (termites or protozoa). Some other bacteria can transform ammonia to nitrate, and nitrate to nitrogen gases, and in that form, it is available to life forms. The degradation of organic material by bacteria and fungi releases fixed nitrogen for reuse by other organisms, thus maintaining the important notion about the web of life (Table **1**).

**Table 1:** The table below shows some estimates of the amount of nitrogen fixed on a global scale (Data from various sources, compiled by Bezdicek and Kennedy [1].

| Types of fixation | $N_2$ fixed ($10^{12}$ g per year) or $10^6$ metric tons per year |
|---|---|
| **Non- biological** | |
| Industrial | about 50 |
| Combustion | about 20 |
| Lightning | about 10 |
| **Total** | about 80 |
| Biological Agricultural land | about 90 |
| Forest and non-agricultural land | about 50 |
| Sea | about 35 |
| **Total** | about 175 |

Legumes such as different species of clover, peas, or beans have root nodules that are the site of nitrogen-fixing symbioses. Microorganism *Rhizobium* is a common example. Each nodule is about 2-3 mm long. Some nodules can be colored. For example, clover root nodules might have a pink color because of the presence of the special pigment leghaemoglobin. It has a unique quality to be synthesized only as the result of bacterial-plant symbiosis. Neither microorganism nor plant alone can produce this pigment. In these leguminous associations, the bacteria usually are *Rhizobium* species, but the root nodules of soybeans, chickpea and some other legumes are formed by small-celled rhizobia. Nodules on some tropical leguminous plants are formed by yet other genera. In all cases the bacteria "invade" the plant and cause the formation of a nodule by inducing localized proliferation of the plant host cells.

Biological nitrogen fixation is performed exclusively by prokaryotes (bacteria and related organisms), using an enzyme complex termed nitrogenase. The reactions occur while N is bound to the nitrogenase enzyme complex. The Fe protein found in the enzyme is first reduced by electrons donated by ferredoxin. Depending on the type of microorganism, the reduced ferredoxin, which supplies electrons for this process, is generated by photosynthesis, respiration or fermentation. These are all the ways for microorganisms to generate energy and convert substances in particular nitrogen.

All the nitrogen-fixing organisms are prokaryotes (bacteria). Some of them live independently of other organisms - the so-called free-living nitrogen-fixing bacteria. Others live in intimate symbiotic associations with plants or with other organisms (e.g. protozoa). *Frankia* is a genus of the bacterial group termed Actinomycetes that are noted for their production of air-borne spores. Included in this group, are the common soil-dwelling *Streptomyces* species which produce many of the antibiotics used in medicine. *Frankia* species are slow-growing in culture, and require specialized media, suggesting that they are specialized symbionts. They form nitrogen-fixing root nodules (sometimes called actinorhizae) with several woody plants of different families. Such as alder (*Alnus sp.*), sea buckthorn (*Hippophae rhamnoides)*, which is common in sand-dune environments) and *Casuarina* (a Mediterranean tree genus). Alder and the other woody hosts of *Frankia* are typical pioneer species that invade nutrient-poor soils. These plants probably benefit from the nitrogen-fixing association, while supplying the bacterial symbiont with photosynthetic products.

The photosynthetic Cyanobacteria often live as free-living organisms in pioneer habitats such as desert soils or as symbionts with lichens in other pioneer habitats. They also form symbiotic associations with other organisms such as the water fern, *Azolla*, and cycads. The association with *Azolla*, where Cyanobacteria (*Anabaena azollae*) are harbored in the leaves, has sometimes been shown to be important for nitrogen inputs in rice paddies, especially if the fern is allowed to grow and then ploughed into the soil to release nitrogen before the rice crop is sown.

In addition to these intimate and specialized symbiotic associations, there are several free-living nitrogen-fixing bacteria that grow in close association with plants. For example, *Azospirillum* species have been shown to fix nitrogen when growing in the root zone (rhizosphere) of tropical grasses, and even of maize plants in field conditions. Similarly, *Azotobacter* species can fix nitrogen in the rhizosphere of several plants. In both cases, the bacteria grow at the expense of sugars and other nutrients that leak from the roots. However, these bacteria can make only a small contribution to the nitrogen nutrition of the plant, because nitrogen-fixation is an energy-expensive process, and large amounts of organic nutrients are not continuously available to microbes in the rhizosphere. This limitation may not apply to the bacteria that live in root nodules or other intimate symbiotic associations with plants. It has been estimated that nitrogen fixation in the nodules of clover roots or other leguminous plants may consume as much as 20% of the total photosynthate.

## NITRIFICATION

The term nitrification refers to the conversion of ammonium to nitrate (pathway 3-4). This is brought about by the nitrifying bacteria, which are specialized to gain their energy by oxidizing ammonium, while using $CO_2$ as their source of carbon to synthesize organic compounds. Organisms of this sort are termed chemoautotrophs. The nitrifying bacteria are found in most soils and waters of moderate pH, but are not active in highly acidic soils. They almost always are found as mixed species communities (termed consortia) because some of them e.g. *Nitrosomonas* species are specialized to convert ammonium to nitrite while others e.g. *Nitrobacter* species convert nitrite to nitrate. In fact, the accumulation of nitrite inhibits *Nitrosomonas,* so it depends on *Nitrobacter* to convert this to nitrate, whereas *Nitrobacter* depends on *Nitrosomonas* to generate nitrite.

The nitrifying bacteria have some important environmental consequences, because they are so common that most of the ammonium in oxygenated soil or natural waters is readily converted to nitrate. However, the process of nitrification has some undesirable consequences. The ammonium ion has a positive charge and so is readily adsorbed onto the negatively charged clay colloids and soil organic matter, preventing it from being washed out of the soil by rainfall. In contrast, the negatively charged nitrate ion is not held on soil particles and so can be washed down the soil profile - the process termed leaching. In this way, valuable nitrogen can be lost from the soil, reducing the soil fertility. The nitrates can then accumulate in groundwater, and ultimately in drinking water. There are strict regulations governing the amount of nitrate that can be present in drinking water, because nitrates can be reduced to highly reactive nitrites by microorganisms in the anaerobic conditions of the gut. Nitrites are absorbed from the gut and bind to hemoglobin, reducing its oxygen-carrying capacity. In young babies this can lead to respiratory distress -

the condition known as "blue baby syndrome". Nitrite in the gut also can react with amino compounds, forming highly carcinogenic nitrosamines.

## DENITRIFICATION

Denitrification refers to the process in which nitrate is converted to gaseous compounds (nitric oxide, nitrous oxide and gaseous nitrogen by microorganisms. The sequence usually involves the production of nitrite. Several types of bacteria perform this conversion when growing on organic matter in anaerobic conditions. Because of the lack of oxygen for normal aerobic respiration, they use nitrate in place of oxygen as the terminal electron acceptor. Thus, the conditions in which we find denitrifying organisms are characterized by (1) a supply of oxidizable organic matter, and (2) absence of oxygen but availability of reducible nitrogen sources. A mixture of gaseous nitrogen products is often produced because of the stepwise use of nitrate, nitrite, nitric oxide and nitrous oxide as electron acceptors in anaerobic respiration. The common denitrifying bacteria include several species of *Pseudomonas*, alkali genes and *Bacillus*. Their activities result in substantial losses of nitrogen into the atmosphere, roughly balancing the amount of nitrogen fixation that occurs each year.

## GENETICAL AND SOME EVOLUTIONARY ASPECTS OF NITROGEN FIXATION

It is clear that the study of bacterial genetics is somehow simpler than for eukaryotes as they lack sexual reproduction and follow the path of binary fission. Bacterial chromosome is circular, not linear. Many prokaryotes have plasmids that are known for fast transfer of genetic material between species promoting drug resistance, toxin production, and more.

The set of genes carrying the genetic information which enables bacteria to fix nitrogen is named nif by microbial geneticists. Nif can come with both positive and negative signs. A minus sign: a mutant with mutated nif, which can no longer fix nitrogen, is called nif-. In more technical terms, nif+ and nif- refer to wild type and mutant genes specifically. Though classical nif- mutants of *Azotobacter* and *Clostridium* had been researched already in the 1960s, the genetics of nif really started with some gene transfer experiments in the 1970s. At this time nif- mutants of *Klebsiella pneumoniae* were obtained and, by conventional methods of gene transfer, nif from wild type *K. pneumoniae* was introduced into the nif- mutants and their mutations corrected. From being nif- they became nif+. The two procedures used to make those gene transfers are transduction and conjugation.

The important conclusion is that nif genes, when present, come under a diligent control system and is related to metabolic aspects. Later recombinant DNA technology has given a means of studying the genetics of almost any diazotroph. Evolution plays its role as well.

As far as the symbioses are concerned, the nodular diazotrophic symbioses among the flowering plants or angiosperms fall into two large groups: Legumes and trees and shrubs. In the 1990s molecular genetic studies revealed that they are more closely related than was thought before.

The nodular symbioses most likely arose from casual associations of plants with certain free-living diazotrophic bacteria which were ancestral to their present day endophytes. In the case of the rhizobia, a most interesting discovery has been the fact that the sym and associated genes are on plasmids. These genetic elements could be transferred between species far more readily than could chromosomal genes during evolutionary time. Authors also discuss the chromosomal background of *Rhizobium* species as that is related to the genus *Agrobacterium* species which cause crown gall growths on susceptible plants. The genetic determinants for crown gall pathogenicity are also plasmid-borne. So it is tempting to regard the two bacterial genera as respectively mutualistic and pathogenic descendants of a common plant-associated (commensal or pathogenic) ancestor. The legumes originated rather early in the evolutionary history of the flowering plants, around 200 million years ago. Nodules are too soft to have left recognizable traces in the fossil record, but it is fair to assume that rhizobial symbioses emerged later than this. Some authorities consider that uneven distribution of nodulation among the three subfamilies of legumes (it is found in some 85% of *Papilionoidae*, 23% of *Mimosoidae* and few *Caesalpinoidae*) implies that rhizobia originated after these groups had diverged, less than 100 million years ago. Such ages are relatively recent in terms of the billions of years of bacterial evolution.

The ways in which bacteria, especially aerobes such as *Azotobacter* and heterocystous *Cyanobacteria*, developed means of utilizing the highly oxygen-sensitive nitrogenase in the presence of air presents yet another fascinating topic in physiological evolution. However, perhaps the most fundamental evolutionary question concerns the origin of diazotrophy itself. Although there is very little paleological evidence to go on, some instructive speculations are possible. Thus, for many years scientists inclined to the view that nitrogen fixation was an ancient property, one which emerged soon after, if not along with, the origin of bacteria themselves, over 3 billion years ago. This opinion arose largely because diazotrophy is only found among the most primitive life forms. In contrast, the cyanobacterial symbioses must surely have originated independently of the nodular symbioses, and probably much earlier because their plant hosts span the whole range of complexity, from primitive to sophisticate.

## REFERENCE

[1]    Bezdicek DF, Kennedy AC. Microorganisms in Action (Lynch JM and Hobbie JE, Eds. Blackwell Scientific Publications. 1998.

**CHAPTER 8**

# Five Years Exploration of Fabricated Soils: Bacterial and Mineral Composition

**Abstract:** The components of FS were changed during the seasons. Nitrogen compounds were reduced in FS during 5 years. More stable were P. K and Ca. Still FS is a good substrate for landscape rehabilitation. The level of macro elements beside nitrogen could be restored by additional application of ash or mineral fertilizers.

## INTRODUCTION

Fabricated soil (FS) was created 5 years ago for the rehabilitation of the mining of soil in Jennings Environmental Educational Center, Butler County, Pennsylvania. The recipe of this soil contains parts of top soil, old mushroom compost, pond sediment, dry leaves and saw dust. This composition makes a favorable ratio of main nutrient elements of nitrogen, phosphorus and potassium (NPK) as well as a sufficient ratio of carbon and nitrogen (1:5). Microbial composition of FS consists of fungi imperfecti and various bacteria species. During the five years of experimentation, trees of poplar and willows were grown on the FS. The level of fungi was diminished and the presence of bacteria increased or varied over the years of research. Five years of investigating FS showed the decrease of nitrogen level, but phosphorus and potassium were not changed actively. The addition of clay in the form of pond sediment increased the level of Al and Si in the FS composition.

FS is a mixture of decaying substrates rich in aluminosilicate, carbon, nitrogen, phosphorus, and potassium sources. This substrate usually is used for landscape rehabilitation. FS are developed from a mixture of materials which encourage plant development. FS in this study was developed for use in the reclamation of drastically disturbed lands. One of the main components of native soil as well as FS is the aluminosilicate matrix provided by clays (illite, smectite, kaolinite, etc.) formed as weathering products of such minerals as orthoclase and other feldspars, and micas, such as muscovite (high potassium content), biotite, and others. These minerals are necessary contributors of calcium, magnesium, sodium, and iron [1]. The size (soil texture) of the mineral fraction in native soil varies from clay-size to coarse sand-size. The carbon and nitrogen-rich organic matter contains the monomers and polymers, the main constituents of the humus complex. Sources of cellulose are dry leaves and sawdust, which also provide lignin. These constituents are humus precursors. Humus is more or less a stable fraction of soil organic matter. It sorbs mineral nutritive elements - nitrogen, potassium and phosphorus, which are important for plant growth and development. Natural soils are commonly described through soil profiles.

Soil is a necessary intermediate substrate in the regulation of the biosphere activity. The loss of soil resources increased up to 10-15 million hectares in a year. Therefore, rehabilitation of the soil cover is a global problem that could be solved by the cooperation of such disciplines as mineralogy, soil science, biology, ecology, agrochemistry, and biochemistry. Developing public and private partnership efforts to utilize (recycle) local waste is promising both with economic and environmental benefits.

The proposed recipes of fabricated soils are based on the concept of a carbon-nitrogen balance in the soil as well as on the transformation of carbon products such as glucose, phenolics, and plant polymers - cellulose and lignin in the humus; polymer that is tightly connected to the aluminosilicate matrix of the soil micelle.

The role of microorganisms in the composting process is important. They combine their activity with plants in the transformation of plant organic substances.

Soil pH is another important soil property that affects the availability of nutrients.

- Macronutrients tend to be less available in soils with low pH.
- Micronutrients tend to be less availablein soils with high pH.

In pH, the desired range of 6.0 to 6.5, nutrients are more readily available to plants, and the microbial populations in the soil increase. Microbes convert nitrogen and sulfur to forms that plants can use.

Macronutrients can be broken into two more groups: Primary and secondary nutrients. The primary nutrients are nitrogen (N), phosphorus (P), and potassium (K). These major nutrients are usually lacking from the soil first because plants use large amounts for their growth and survival.

The secondary nutrients are calcium (Ca), magnesium (Mg), and sulfur (S). There are usually enough of these nutrients in the soil so fertilization is not always needed. Also, large amounts of calcium and magnesium are added when lime is applied to acidic soils. Sulfur is usually found in sufficient amounts from the slow decomposition of soil organic matter, an important reason for not throwing out grass clippings and leaves.

Micronutrients are boron (B), copper (Cu), iron (Fe), chloride (Cl), manganese (Mn), molybdenum (Mo), and zinc (Zn). Recycling organic matter such as grass clippings and tree leaves is an excellent way of providing micronutrients (as well as macronutrients) to growing plants.

Soil sampling was done on the FS plots. The starting year FS-2002 was used as a control. Other samples were used for the investigation of mineral composition and plant and microbiological activity of the FS. Mineral elements were determined by Conti, Inc. USA. Microbial analytics was done by US Microsolutions, Inc.

## RESULTS

FS contains two forms of carbon sources- dry leaves and saw dust. Leaves subjected to composting in 2 months and saw dust carbon transformed in 2-3 years. FS contains more carbon than regular Gresham soil (Table **1**).

**Table 1:** Legend list of the investigated soils: Macro and micro-elements in mining and fabricated soil (FS) 2002-2007.

| Number | Location | Sample |
|--------|----------|--------|
| 1 | | Mining soil-2002 |
| 2 | Left | FS under willow plants 2007 left |
| 3 | right | FS under willow plants- 2007 right |
| 4 | Center | FS under willow plants- 2007 middle |
| 5 | 3 Left | FS near willow plants-2007 left |
| 6 | 3 Right | FS near willow plants-2007 right |
| 7 | 3 Medium | FS near willow plants-2007 middle |
| 8 | 4 | FS-2002 |
| 9 | 5 | Mining soil-2007 |

After one year, FS contains almost the same carbon on the willow and poplar plots. The levels of mineral elements after one year of FS exposure on the mining soil were much higher than in Gresham soil. Gresham soil is considered standard and is nearly level, very deep, somewhat poorly drained. Typically the surface layer is dark grayish brown silt loam about 8 inches thick (Soil Survey of Butler County, 1989).

Thus one year of FS exposure in nature did not change the level of nutritive elements. This was accompanied by fast growth of willow and poplar plants. The amount of nitrogen increased the most during 2002-2003 of FS exposure. The increase in nitrogen concentration might be the result of active composting processes that took place in manufactured soils. Definitely the combination of compost effects and decay of leaf materials contributed to the increase of the nitrogen concentration specifically within the first three months of soil exposure. At the same time, the total bacterial count and number of bacterial species increased in FS during one year of the exposure. However, mining soil was very poor in microbial cenosis (Table **2**).

**Table 2:** Number of bacterial colonies found in soil samples.

| Soil Type | Total bacterial count CFU x $10^6$/number of bacterial species | |
|---|---|---|
| | **2004** | **2005** |
| Mining soil | .0022/8 | 0.063 |
| Top soil | .43/8 | 711 |
| Fabricated soil | 27/9 | 711 |
| Fabricated soil 2002 polar | 66/10 | 68.4 |
| Fabricated soil 2002 willow | 18/8 | 486 |
| Fabricated soil 2003 polar | 114/14 | 495 |
| Fabricated soil 2003 willow | 30/8 | 882 |

*All data is stataistically significant. 95% of confidence intervals exist for all points, a= 0.05.

In general, it is important to mention that FS during one year of exposure was still rich in nutritive elements, with the favorite carbon-nitrogen content on both types of plots-willow and poplar. However, after five years of exposure on the some lots, FS components have been changed (Table **3**).

**Table 3:** Reduce of elements presence in % in FS after 5 years of exposure on willow plot.

| Elements | N | P | Ca | K | Mg | Mn | Fe | Cu |
|---|---|---|---|---|---|---|---|---|
| Change in % (2007-2002) | 30 | 77 | 88 | 64 | 25 | 25 | 34 | 25 |

In 2007 (since 2002) the most intensive loss of nutritive elements in FS was observed in nitrogen and some microelements, less for calcium, phosphorus and potassium. That means that during intensive growth of willow plants, the nitrogen is utilized faster than other elements. The comparison of FS plots under chestnut and willow showed that the total bacterial count was similar, but it appeared that under the willow there were new forms of microorganisms found: *Sphingomonas, Chryzeobacterium* and *Rhizobium.*

Microbial content in FS (seed plot, chestnut project, and willow mining soil) contains mostly Actinomycetes, and the amount of bacterial species was more than two times reduced.

Biological activity of different soil components was also determined by using four plant crops sensitive to the presence of bioactive compounds (Table **4**).

**Table 4:** Determination of biological activity of different soils with willow and chestnut plants.

| Number | Samples | pH | Crops (shoot length in cm) | | | |
|---|---|---|---|---|---|---|
| | | | **Turnip** | **Lettuce** | **Rye** | **Clover** |
| 1 | Forest soil | 6.5 | 5.2 | 4.1 | 8.1 | 4.7 |
| | **Chesnut Project one** | | | | | |
| 2 | Forest soil | 6.4 | 4.6 | 3.5 | 7.9 | 4.0 |
| | **Chesnut Project two** | | | | | |
| | Average Forest soil | | 4.9= 100 % | 3.8= 100% | 8.0= 100% | 4.4= 100% |
| 3 | De Sale Front | 6 | 6.5 | 3.1 | 8.9 | 4.3 |
| % | | | 102 | 98 | 111 | 97 |
| 4 | De Sale Front | 6.7 | 4.6 | 3.3 | 7.6 | 5.0 |
| % | | | 94 | 87 | 95 | 114 |
| 5 | FS Seed plot | 6.8 | 4.2 | 3.0 | 8.2 | 3.6 |
| | **Chesnut Project** | | | | | |
| 6 | FS Trees plot | 6.8 | 4.9 | 2.9 | 7.3 | 4.5 |
| % | | | 100 | 76 | 91 | 102 |
| 7 | De Sale Mn | 6.7 | 4.8 | 3.0 | 7.4 | 4.8 |
| % | | | 98 | 79 | 93 | 109 |

All tested forms of the soils: Forest soil (Gresham type), and De Sale soil (Ernest series), had neutral pH and more or less high biological activity. This form of testing should be considered in conjunction with woody plant growth as an integrative test for soil characteristics. Thus, FS was used for landscape rehabilitation and the growth of hard wood plants: willow, polar and chestnut trees. Willow plants excreted such allelopathic substances as salicylic acid that had anti-inflammatory activity and protected chestnuts from blight. Chestnut trees are prone to blight infection.

The complex investigation of the FS showed that the mineral elements composition and biological characteristics of the soil samples, including microbacterial analysis, showed the important ecological role of FS for landscape restoration.

## REFERENCE

[1]     Brady NC. The Nature and Properties of Soils. 3rd Edition, MacMillan Publ., New York. 1994; pp. 604.

# CHAPTER 9

# Carbon Dioxide, Photosynthesis, Growth, and Productivity

**Abstract:** Solar energy as a renewable form is a key source of photosynthesis and plant biomass accumulation. The composition of plant biomass is based on the formation of primary and secondary products of photosynthesis. Some of them are carbon containing polymers like starch, cellulose and lignin. These components of plant organism could convert to soil organic matter (humus and its components).

## INTRODUCTION

Leaves and stems of plants are able to develop their body thanks to the main gas exchange processes: Photosynthesis and respiration. Leaves are the centers of the biosynthesis of plant monomers (glucose, phenol acids and alcohols, amino acids, nitrogen bases) from carbon dioxide in the air. Monomers are blocks of plant polymers which construct the body of the plant. Thanks to polymers of plant origin such as proteins, starch, cellulose, and lignin, the self-developing construction of the plant body could grow and become more resistant to environmental factors [1].

Conducting vessels in the bark and wood of the plants are the transporting arteries for the movement of monomers and polymers from leaves and roots. The pipes and tubes in the plant structure are also a real network, called the armature, of the plant body. The centers of the regulation of plant activity, plant hormones and inhibitors (growth regulating substances), are mostly located in the growing points of stems and roots. Growing points or apexes act as the centers for the regulation of the growth and self-reproduction activity.

Compartmentalization (location in special compartments) of the processes of biosynthesis and the deposit of the storage substances is a result of transport of substances over long distances. All these processes are based on photosynthesis, growth and regeneration of plant organs (rhythms of activity and dormancy). The formation of the yield (plant productivity) is a result of this complex machinery function. Plants could be considered as Phyto-Architecton or the self-developing architectural model with the centers of body formation, operation of the activity, conducting vessels and the deposit of the accumulated products [2].

The Phyto-Architecton is based on the formation of cellular and organ blocks, which are used for the formation of renewable forms of energy. At the same time, Phyto-Architecton controls its body building thanks to light and geo-sensors as well as to hormone biosynthesis which directs the nutrition flow to the proper compartments.

## CARBON DIOXIDE AND PHOTOSYNTHESIS

Carbon dioxide is one of three main components which combine to produce the products necessary for plant development. The amount of carbon dioxide ($CO_2$) in the air is only 0.003%. This compares to 78% of nitrogen, 21% of oxygen, and 0.97% of trace gases. Plants will stop growing when $CO_2$ decreases below 150 ppm. A higher $CO_2$ concentration' (400-500 ppm) increases the rate of formation of dry plant matter (Fig. **1**).

## C-3 AND C-4 PLANTS

$CO_2$ /$O_2$ relations are based on two global processes, photosynthesis and respiration. There are 3 main types of plants in their relation to respiration. C-3 plants such as cucumbers, tomatoes, and carnations evaluate $CO_2$ during respiration during the night. C-4 plants conserve metabolic $CO_2$ in organic forms (mostly acids of the respiration cycle). The main C-4 plants are bamboo, rice, corn, sugar cane, millet, and orchids. CAM-plants have the other type of $CO_2$ conservation. C-4 plants possess higher productivity of biomass and do not produce much $CO_2$ as the product of glucose splitting.

## $CO_2$ IN A GREENHOUSE

Usually the level of $CO_2$ in a green house is a limiting factor for plant growth and harvest. The $CO_2$ concentration in the center part of a green house is lower than near the outer walls. The lack of adequate $CO_2$ levels lowers the average plant yield quality and market value. Growers using a 50% higher $CO_2$ concentration in a green house increased crop production without costly methods of stimulating plant growth.

## CROP SPACING

A rule of thumb for soil and bag culture is that each plant needs a 0.4 square meter growing area and that total plant density should be 24.710 plants per hectare. Generally, tomatoes in plant bag culture are set out in double rows. The germination rate for greenhouse tomatoes is generally at least 80%, so only one seed needs to be planted per container. Plant shape depends on branching intensity and growth, which are controlled by genetics and selection properties. Dwarf and semi dwarf plants have retarded growth and higher productivity. Thus the choice of plant varieties is an important factor of green house plant productivity (green mass of leaves and stems plus harvest (fruits).

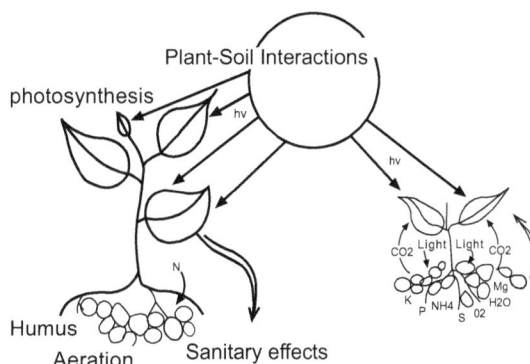

**Figure 1:** Plant-soil interactions. Solar energy transformed to energy of soil fertility via photosynthesis. At the same time exist gas, ion and organic product exchange between plant and soil.

Effects of UV-B and UV-C on the seedlings of some crops were studied in conjunction with the growth of roots of irradiated plants resistant to both types of UV-light (30000 J/sq.cm) like pea, bean and wheat. Lettuce and *Arabidopsis* seedlings are very sensitive. The growth of roots of irradiated *Arabidopsis* was inhibited by 50% if the dose 3000J/sq.cm was applied. As a rule, roots were more sensitive to UV than shoots. One of the primary effects of UV-C on the seedlings was activation of potassium and sodium excretions. Calcium was more stable and did not secrete actively. Dry seeds were more stable to UV stress, than germinated seeds or seedlings. *Arabidopsis* mutants with different sensitivity to UV-C were also tested. Level of ethylene and abscisic acid was the highest in the UV-B resistant mutants [3].

Solar energy is transformed to energy of soil fertility *via* photosynthesis. At the same time, there exists gas, ion, and organic product exchanges between plants and the soil. Humus formation is the base of plant productivity. In this process, mineral and organic elements of the soil are participants, as are microbial factors.

## REFERENCES

[1]    Kefeli VI and Kalevitch MV, Potasova NN. Growth, photosynthesis and mineral nutrition. Physiology and biochemsitry of cultivated plants. 1992; 1: 64-82.

[2]    Allakhverdiev SL, Muzafarov EN, Klimov V. Quercetin effect on electron transfer in photosystems I and II of pea chloroplasts. Biofizika, 1989; 34: 976-979.

[3]    Kefeli VI, Shotweli M, Banko A, Vlasov P, Rakitina T. UV light, hormones and plant growth. The 11th Congress FESPP. Bulgarian J of Plant Physiology Special Issue. 1998; Abs. 328.

# CHAPTER 10

# Chemical Signaling During an Organism's Growth and Development

**Abstract:** The accumulation of plant biomass in plants is regulated by two factors – genomes and plant growth regulating system. Hormones as growth regulating factors are located in different zones of plant organisms. They have specific effects in plant bodies like auxins (roots induction), gibberellins (stem elongation), cytokinins (cell multiplications).

## INTRODUCTION

Only a few researchers have studied, correlative processes in multicellular organisms (Kholodny, Kurssanov and Chailakhjan). However, practical applications of scientific research and a holistic approach itself are tightly connected with correlative messages based on genetic, ionic, and hormonal integrative systems. These systems do determine growth and development of the organisms. All correlative signals could be divided into internal and external signals. Among external signals one can find physical signals such as temperature, light, gravity, magnetic fields, etc. Internal signals are chemical signals such as mineral elements, organic substances, food components, phytohormones and others (Fig. **1**).

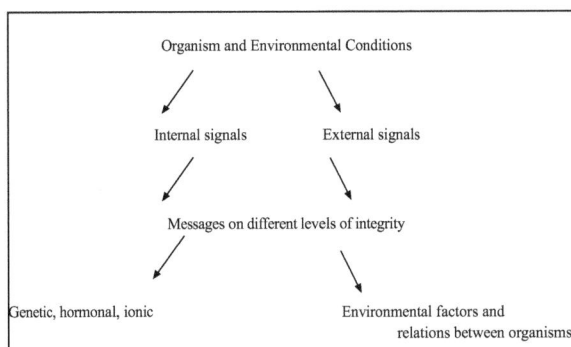

**Figure 1:** Correlations and signals in living entity.

Different living organisms such as plants, microorganisms, and certainly, animals and humans are participants in this signaling exchange. Environmental effects are also taken into consideration [1-3] Correlations could be connected and disconnected, as it was previously shown on the models of plant growth and organogenesis [4].

The concept of correlations between hormones and anti-hormones in plant tissue was chosen for this study. Even though it is not a new concept, it is still yielding great potential and was chosen by authors for detailed investigation.

Previously this concept was discussed by Kefeli and Kadyrov [5] and a few years later, Leopold and Kriedemann [6] suggested that this scientific idea of signaling is mostly based on the early interrelations between biosynthesis of phenolic and indolic substances, rather than on the quantity of these naturally occurring plant chemicals. Molecular interactions between hormones and antihormones could not be the only basis for morphological correlations, but also for the intermediate messages under the effect of the external factors.

Thus, Cotuk [7] in his review of the book by Ozalpan, 'Basic Radiobiology' wrote that for understanding the biological effects of radiation on the organism, it is necessary to summarize the effects on biomolecules, subcellular structures, cells, tissues, and organs respectively. Earlier Tretjakov and co-workers [11] while continuing the traditions of Maximov (Honorary Member of the American Society of Plant Physiologists) also discussed the similar concept of phytohormone and inhibitor interactions during plant growth. He confirmed our idea that the growth process is a balance between plant hormones and natural growth inhibitors. These relations are modified during the life cycle of the plant. Our latest data helped us to contribute to the development of the concept that deals with the role of plant hormones and inhibitors in the life of single plant and plant cenosis.

**Narcin Palavan-Unsal (Ed)**

Some chemicals found in abundance in plant tissue are phenolics. Their main properties include: presence in tissue is easily determined; they participate in oxidative reactions in the cells; and they are not transported to long distances from the point of formation as opposed to hormones which are transported long distances. Phenolics can be the main bioactive compounds of decaying humus in manufactured soils. They are also found in many plant species including willow trees. This internal system of chemical signals is the only part of endogenous systems of regulation. We continue to develop this concept and research further the effects of ecological factors on plants and their internal signaling processes.

## ROLE OF PHENOLICS IN PLANTS

Willow leaves contained coumarins and phenolic acids, few flavonoids and no anthocyanins. Hydroxy-coumarins like umbeliferone (UF) could be used as a standard for the determination of the quantity of phenolics in water extracts. The different concentrations of UF on chromatograms formed pigments or colorless stains with different square areas. The higher dosage of UF apparently made a larger stain or pigmentation. With the help of diazosulfanilic acid as a color reagent it was possible to observe UF pigmentation and then create a calibration curve.

The more specific substance for the willow species that were under investigation was flavonol with the Rf not higher than 0.2. The higher amount of flavonol found in *Salix viminali,* less in *Salix glaucophylla.* All three investigated species contained coumarins with the Rf 0.8, a bright blue color reaction. Paper chromatography and color reactions could be helpful tools in identification of chemical markers. Thus this method of chemo- taxonomy could be a sufficient supplement for establishing real identification of some species with proper morphological features.

Other tree species such as sumac leaves, for example, contained flavonoids which had low mobility and were yellow in color under UV light. The application of baking soda spray increased its intensity giving a pink color in daylight from anthocyanins, blue in the UV from coumarins, and changed color to green with baking soda, Rf 0.58. Our previous research also supported these data.

## INTERACTIONS BETWEEN PLANTS AND ENVIRONMENT

Some substances could be retained by willow roots and some are excreted in the external water medium. Chromatography of these water exudates and the subsequent investigation of chromatograms in UV-B light showed that most of these substances are polyphenols, coumarins, or phenolic acids. The phenolic substances which are retained by the cells have different properties than the ones that are located in the root exudates. These data confirm the idea about excreted substances being allelopathogens in nature and their involvement in ecological interactions between different plant species.

## ENVIRONMENTAL FACTORS AND PLANT REACTIONS

The effects of UV-B (280-320 nm) and UV-C (300-354 nm) on the seedlings of some crops were also studied. There are some plants that are resistant to both types of UV -light (30, 000 J/sq. cm) like pea, bean or wheat. Lettuce and *Arabidopsis* seedlings are very sensitive. The growth of roots of irradiated *Arabidopsis* seedlings was inhibited 50% (3000 J/sq.cm was applied). If the intensity of solar light increased, the petioles of leaves become shorter (effect of photomorphogenesis). It was observed that the leaf blade became thinner and the petiole of the leaf became longer in the case of low light intensity. As a rule, roots were more sensitive to the UV effects than shoots. One of the primary effects of UV-C on the seedlings was activation of potassium and sodium secretions. Calcium was more stable and was not secreted actively. Dry seeds were more stable to UV stress, than germinated seeds or seedlings. *Arabidopsis* mutants with different sensitivity to UV-C were also tested. The levels of ethylene and ABA were the highest in the UV-B resistant mutants. These data correspond to the ideas presented by Ozalpan [3].

## PHENOLICS IN THE LEAVES OF DIFFERENT AGE

The leaves of scarlet maple, *Acer rubrum* L. were collected in July (green), September (abscised, yellow) and in February (brown, composted). Water extracts of these leaves were subjected to paper chromatography in 5% acetic acid. Chromatograms were investigated under UVB light and sprayed by baking soda and Pauli reagent (diazo-sulfanilic acid). This complex of color reactions allowed identifying monophenols-coumarins and phenol-carbonic acids as well as flavonoids and anthocyanins. During the composting process, the pattern of phenolics in leaves had changed remarkably. The most dramatic changes were observed in the flavonoid complex. Incubation of water extracts with fungi imperfecti changed the phenolic complex; the amount of anthocyanin was decreased. The concept of the interaction between the active molecules during ontogenesis is accepted by Palavan-Unsal [2].

## TRANSFORMATION OF PHENOLIC COMPOUNDS IN MAPLE LEAVES DURING COMPOSTING

Leaves are a good source of carbon in the composting process. During the microbial decay of leaves, some chemicals in the soil split very quickly, and some are more stable. Among stable substances are phenolics with some inhibitory activity.

To determine pH of leaves (water extracts) and biological activity of these extracts, the following experiments were done and results obtained:

1.  The most specific substance for the willow species investigation was flavonol with the Rf not higher than 0.2.

2.  A higher amount of flavonol possessed by *Salix viminalis*; lesser by *Salix glaucophylla.*

3.  All 3 investigated species contain coumarins with the Rf 0.8, a bright blue color reaction.

It is interesting to mention that leaves during composting lost their acid reaction and their inhibitory activity. These effects made maple leaves and sumac leaves more acceptable in an ecological sense. Leaves during 1-4 months also lost their allelopathic activity and became just a carbon substrate for the composting process.

## AGE AND REGENERATION

Willow (*Salix discolor* Muhl) and poplar (*Populus nigra* L) cuttings were obtained from one year shoots of the original plants. Cuttings were equal in their length, 18 cm, but differed in their weight remarkably. Cuttings from the lower part of the shoot were 3 times heavier for both plants than cuttings from the upper part of the shoot. The weight of the leaves on the corresponding parts of the willow mother shoots was higher in the central part of the shoot and lower in the upper and basal parts of the shoot. Poplar cuttings from the upper part of the shoot formed less roots and shorter roots than cuttings from the lower part which also formed shorter roots. We suppose that the age of the shoot had an influence on the rooting process not only *via* hormonal balance but also *via* the biosynthesis of carbon polymers such as lignin and cellulose.

## BIOASSAYS IN DETERMINATION OF PLANT GROWTH ACTIVITY

Biological activity of polluted water, contaminated or healthy soil, water extracts and root exudates could be analyzed by a proposed complex of seed bioassays. Lettuce, wheat, mustard and clover are plants from different families which can easily germinate and have no effect on the growth of each other. Germination of seeds of these test plants as well as growth of their stems was important criteria in the determination of toxic and inhibitory activity of the substrates. Because of that seed property it was possible to investigate the substrate in one Petri dish separated into quadrants. The test itself is very rapid and data could be obtained after one week. This test is successfully applied to both the investigation of fabricated soils (artificial soils created for revitalizing eroded soils) and exudates from chromatograms.

The roots and leaves of willow and poplar contained salicilate, coumarins and some phenolic acids. Part of these substances could be secreted from roots and could be subjected to microbial transformation.

The cuttings of willow (*Salix excelsior*) and black poplar (*Populus nigra*) were rooted in water and the rooting process was investigated. The upper part of the mother shoot is usually less active for rooting in comparison with the basal part. After rooting, the cuttings were planted into FS plots.

## PHENOLIC SECRETIONS AND LEAF-INHIBITORY PROPERTIES

Natural plant exudates may often inhibit seed germination and seedling development, thus triggering allelopathy within soil systems and other growing media [8]. The purpose was to determine the ability of mustard seeds to recover from an extended exposure to various extracts from sumac leaves (*Rhus typhina*) which excrete allelopathic compounds that inhibit germination and growth of some types of plants when these accumulate in adjacent rhizospheres.

Four different types of seeds of common cultivated agronomic crops were placed into Petri dishes lined with filter paper, which was divided into quadrants. The filter papers were imbibed with 2 ml each of sumac leaf and root extracts, while tap water served as a control. Tested in each of the quadrants were mustard, red clover, lettuce and wheat, of which mustard appeared with no germination. The seeds were then removed from testing conditions, rinsed in tap water, and placed back into control conditions in order to study their capability to grow. After 48 hours under these conditions, 80% of mustard seeds germinated without apparent inhibition (Tables **1-4**).

These initial observations indicate that the allelopathic exudates of sumac, most particularly those extracted from red, fall leaves, only temporarily stunt the growth of mustard. This shows that the inhibiting substances do not have long term effects on the process of seed germination and that the search for botanical herbicides amongst natural sources, like leaves and roots of sumac, are less harmful than synthetic herbicides. Over time, as the chemical compounds are leached out of the soil, the seeds germinate. Paper chromatography showed that sumac leaves contain at least four different groups of phenolics, which were identified by Rf position, UV light, and soda. Thus, water extracts of leaves can yield at least four classes of phenolic substances such as anthocyanins (red pigment), flavonoids (yellow pigment), coumarins (colorless, bright blue under UV light), and phenolic acids. These substances have different Rfs on chromatograms. The phenolic complexes of the leaves are more toxic than the phenolic complexes of the roots. The phenolic complexes of sumac leaves (collected in autumn) inhibit mustard, clover, and also exert a potent inhibiting effect on lettuce, whereas wheat is most resistant. Therefore, the phenolic complex of leaves may be used as a source for botanical herbicide with inhibiting effects on the germination of weeds that belong to the Cruciferae family. Although this inhibition is not permanent, but temporary, it may suffice if the treatment is properly timed to compete effectively against unwanted vegetation in the cultivated field or nursery and with foreseeable environmental benefits [8].

Table **1** shows the selective inhibiting effect of leaf extracts on the seed germination of the 4 crops. The most sensitive to the extracts were mustard seeds and the extracts from the roots of sumac were inert.

**Table 1:** Effect of water extracts from sumac leaves and roots on the germination of the seeds of 4 crops (in % to control). Dry material to water ratio 7:1.

| Trial/Extract | Wheat | Clover | Mustard | Lettuce |
|---|---|---|---|---|
| Control (water) | 46 | 74 | 78 | 86 |
| Leaves, green | 51 | 40 | 0 | 68 |
| Leaves, red, abscised | 63 | 22 | 1 | 23 |
| Roots | 53 | 93 | 92 | 90 |

These data show that mustard seeds are mostly sensitive to the inhibitory effect of water extracts from the green leaves of sumac. Such selectivity in the reactions is similar to the herbicidal effect. This type of botanical herbicide has beneficial, short term effects (Table **2**). After washing the seeds, they restore the capability to grow.

**Table 2:** Effect of water extracts from sumac leaves and roots* on the seedlings of the 4 crop tests (length of seedlings in mm and in % to the control).

| Trial | Length of the seedlings (mm) | | | | | | | |
|---|---|---|---|---|---|---|---|---|
| | Wheat | | Clover | | Mustard | | Lettuce | |
| | Average | % | Average | % | Average | % | Average | % |
| Control (water) | 99 | 100 | 34 | 100 | 35 | 100 | 34 | 100 |
| Sumac abscised leaves | 29 | 29 | 15 | 44 | 14 | 40 | 20 | 59 |
| Sumac roots | 75 | 75 | 28 | 82 | 28 | 82 | 23 | 68 |

*10 g of dry matter was extracted 24 h by 70 ml of water at 18°C.

Table **3** demonstrates that extracts from sumac leaves are acidic, probably because of the presence of weak phenolic acids such as p-coumaric, caffeic and other C-6, C-3 products. These substances are not present in the roots in significant amounts.

**Table 3:** pH of sumac leaves and root extracts.

| | Trial | | |
|---|---|---|---|
| | **Control** | **Leaf extract** | **Root extract** |
| **pH** | 7.66 | 3.63 | 5.89 |

Previous data show the effects of sumac leaf extracts on the processes of germination and growth. Table **4** demonstrates a strong inhibitory effect of the extract on the rooting of bean cuttings.

**Table.4:** Effect of water extracts of sumac leaves and roots on the rooting processes of bean plants.

| Trial | Number of roots on 1 cuttings | | Length in mm and % to control | | | |
|---|---|---|---|---|---|---|
| | | | Longest root | | Shoot | |
| | **Average** | **%** | **Average** | **%** | **Average** | **%** |
| Control (water) | 18.6 | 100 | 90 | 100 | 63 | 100 |
| Standard rooting hormone, indolyl-3-acetic acid (IAA, 70 mg/l) | 32 | 170 | 310 | 340 | 193 | 300 |
| Sumac leaves | 0.0 | 0.0 | 0.0 | 0.0 | 0.0 | 0.0 |
| Sumac roots | 26.6 | 142 | 302 | 143 | 195 | 300 |

A similar inhibitory effect, but not as strong as before was observed in the case of stem wood cuttings of the willow trees (Table **5**).

**Table 5:** Effect of water extracts of sumac leaves and roots on rooting processes in willow (*Salix discolor* Muhl) stem cuttings (R/C: amount of roots per cutting; C: Control).

| Trial | Willow rooting after | | | | | |
|---|---|---|---|---|---|---|
| | 2 weeks | | 3 weeks | | 4 weeks | |
| | R/C | % to C | R/C | % to C | R/C | % to C |
| Control | 3.5 | 100 | 11.8 | 100 | 17.0 | 100 |
| Leaves | 1.6 | 45 | 6.4 | 54 | 15.0 | 88 |
| Roots | 3.8 | 111 | 9.8 | 83 | 15.0 | 88 |

## INDOLIC AUXINS AS CHEMICAL SIGNALS

The natural auxin, indole-3-acetic acid (lAA) in the concentration 70 mg/l, induced the formation of roots on the stem cuttings of bean plants (*Phaseolus vulgaris* L.) after 12 hours of incubation in the auxin solution and successful rooting in water during 7 days at 22˚C. Water was used in the control and only 4 roots grew, whereas in lAA, 60 roots grew. If IAA was oxidized by the bean root enzyme peroxidase, the auxin effect disappeared (Tables **6-7**). If auxin was applied to the cuttings not at the beginning of the experiment but after root primordial formation (3$^{rd}$ day), the effect of auxin was lost. These data showed that auxin induced rooting *via* some intermediate processes. The pretreatment of the cuttings with the low dosages appreciably attenuated the stimulating effect of lAA on the root formation process of metabolic inhibitors actinomycin D, 8-azaguanine and chloramphenicol. It was shown that root forming cuttings in which cell division occurs 500-1000 times higher are more sensitive to the inhibitors of nucleic acid-protein metabolism compared to coleoptile segment growth. This coleoptile growth is based on cell elongation, and is also sensitive to IAA (7 mg/l). Willow (*Salix discolor* Muhl) stem cuttings after the treatment by lAA, 150 mg/l, formed roots after 14 days of incubation at 22°C [9].

**Table 6:** Effect of indole-3 acetic acid (lAA), 70 mg/l, on the rooting of bean cuttings.

| Trial | Results for 1cuttings | | | | | |
|---|---|---|---|---|---|---|
| | 3 experiments (1 and 2 represent the exposure in weeks) | | | | | |
| | Number of roots | | Length of the longest root (mm) | | Length of shoot (mm) | |
| | 1 | 2 | 1 | 2 | 1 | 2 |
| Control (water) | 2.9 | 15 | 7.4 | 3.3 | 7.7 | 15 |
| IAA, penetration via stem cur | 5.3 | 40 | 7.6 | 35 | 7.6 | 19 |
| IAA, penetration via roots | 1.1 | 32 | 3.0 | 23 | 9.6 | 14 |

**Table 7:** Effect of oxidized lAA on rooting of bean cuttings.

| Trial | Results for per cuttings | | | | | |
|---|---|---|---|---|---|---|
| | 3 experiments (1 and 2 represent the exposure in weeks) | | | | | |
| | Number of roots | | Length of the longest root (mm) | | Length of shoot (mm) | |
| | 1 | 2 | 1 | 2 | 1 | 2 |
| Control (water) | 2.4 | 20.3 | 8.2 | 79.8 | 2.9 | 31.3 |
| IAA (75 mg/l) | 12 | 39.6 | 10.4 | 56.8 | 6.8 | 32.1 |
| IAA preliminary passed through Roots* | 8.6 | 35.2 | 13.6 | 49.0 | 11.8 | 26.8 |
| Oxidized lAA (OIAA)** | 1.2 | 9.0 | 2.3 | 3 | 4.1 | 5.0 |

* Roots as centers of auxin-oxidase, peroxidase (OIAA)
** OIAA-oxidized IAA. Preparation of the peroxidase from bean roots consisted of roots from 10, 2-week old beans plants (3.44g) mixed with IAA, 75 mg/l, in a ratio 1:1 and used for cutting treatment.

If preliminary bark and cambium of the stem cuttings were separated from wood, the rooting process stopped. If only a strip of bark and cambium was separated, rooting proceeded on the intact part of the cutting.

Thus, the rooting process is connected with the bark-cambium complex. This complex is inactive during the autumn dormancy period (1.5-2 months). During this time, buds did not open and roots did not form even in the lab conditions. The rooting process of dormant cutting is suppressed but not as long as bud dormancy. Usually buds attached to the rooting zone open earlier than upper buds of the stem cuttings. These data showed the effect of a local break of dormancy and probably local activation of native auxin biosynthesis in the zone of the stem cambium. We checked the possibility of auxin biosynthesis in the isolated cells of *Dioscorea* and confirmed its possibility. It is necessary to mention that the tagged cells sensitivity of cambium is very important for root initiation. Table **6** shows the strongest effect of IAA while penetrating *via* stem cut, but not *via* roots of bean plants.

IAA stimulated the rooting of cuttings only if it penetrated into the cutting via stem cuts. Preliminary oxidized IAA became inactive (Table **7**). These data are proof that root enzymes, peroxidases, are compartmentalized and separated from xylem pathways.

Chemical signaling in plants proceeded on metabolic, hormonal and ionic levels. Some of the regulating processes are associated with the cell structures and some with tissue or organ development [6]. Plants, as any other living organisms, accept signals from the environment and send signals outside in forms of ions and organic substrates. Among these substances could be relatively stable products like phenolics which can regulate growth and development activity of other species and participate in the construction of the natural cenosis. However, unstable chemicals are also present [4, 10].

Also, it is important to mention the *Fabacea* species - clovers and trefoils - possess the ability of nitrogen fixating by very good visible nodules. These plants could participate in soil nitrogen stabilization [1]. Plants are able to interact by chemical signaling which might be observed on FS during the investigation of weed plant composition on the plots. Allelopathy is a common way of interacting between living components of ecosystems.

Disruption of the signaling system led to the development of abnormalities, such as interruption in processes of tissue differentiation and tumor formation. Reconstruction of plant integrity proceeds with the help of phytohormones, mostly indolic auxins [9, 11]. It is essential to consider that after plant death, its organic matter is subjected to humus transformation. Primary stages of this transformation could be connected with the elaboration of phenolic substances from the cells, and they could play a role of allelopathogens, thus becoming crucially important in the signaling system. This is just an attempt to discuss the issues of chemical signaling in the plant or plant community. Ion regulation of plant organisms under normal and stress conditions is the next step in deciphering the processes of endogenous signaling.

## REFERENCES

[1]    Kalevitch MV, Kefeli VI, Borsari B, Davis J, Bolous G. Final version chemical signaling during organism's growth and development. Journal of Cell and Molecular Biology 2004; 3: 95-102.

[2]    Palavan-Ünsal N. Book Review. Natural Growth Inhibitors in Plant and Environment, Journal of Cell and Molecular Biology 2003; 2: 60.

[3]    Ozalpan A. Basic Radiobiology. Halic University Publ, Istanbul 2001; pp. 353.

[4]    Kefeli V, Kalevitch MV Natural Growth Inhibitors and Phytohormones in Plants and Environment, Kluwer Academic publishers, Dorderecht/Boston/London 2003; pp. 1-322.

[5]    Kefeli V, Kadyrov C. Natural growth inhibitors, their chemical and physiological properties, Annual Review of Plant Physiology 1971; 22: 185-196.

[6]    Leopold C, Kriedeman P. Plant Growth and Development, McGraw-Hill, N-Y, Second Edition, 1975; pp. 539.

[7]    Çotuk Y. Peer Review. Book A. Ozalpan (2002) Basic Radiobiology, Journal of Cell and Molecular Biology, Halic University, Turkey, 1: 39–40.

[8]    Boulos G, Borsari B, Kefeli V. Botanical herbicides in action. Jour of Scholarly Endeavor, Slippery Rock University, 2004; 4: 10.

[9]    Davis J, Borsari B, Kefeli V. Monitoring root formation of willow cuttings under the effects of a selected growth promotion and retardation. J of Scholarly Endeavor, Slippery Rock University, 2004; p. 4: 10.

[10]   Tretjakov NN. Physiology and Biochemistry of Agricultural Plants. Moscow 2000; pp. 639.

[11]   Kefeli V. Researches of Plant Hormones, 1959-2004. Personal Impressions, In: Auxins- Orthodox Academy of Crete, organizers A. Theologis and G. Sendberg, Crete, Greece. Conference Abstracts 2004; 1-70.

# CHAPTER 11

# Phytohormones in Transformed Tobacco Plants

**Abstract:** Soil bacteria *Agrobacterium tumefaciens* and *Agrobacterium rhizogenes* are able to modify plant cell genome and activate the level of plant growth hormones Thus, the gene mutated plant cell cultures could form new types of plants with the higher amount of phyto-hormones. This modification is a result of interaction of soil bacteria with higher plant cells.

## INTRODUCTION

There are still unclear areas in our understanding of the function of the hormonal system in plants. These problems include, for example, such matters as the significance of hormone ratios and interactions between growth stimulating hormones and inhibiting factors of different types, as well as the input of soil bacteria into plant genome construction.

We are approaching this problem by way of experiments with transgenic plants transformed by modified genetic lines of *Agrobacterium tumefaciens*. In these plants, hormonal levels may be altered. Transgenic plants could be produced by genetic engineering procedures. Foreign genes are inserted into the host genome using different vectors. One very useful vector is the Ti plasmid of *Agrobacterium tumefaciens*.

In the last few years, transgenic plants with integrated bacterial gene 4 have been obtained by this procedure. It was shown that regenerated plants with transferred genes 4 are dwarfed and root with difficulty. The authors of these studies used, for the transfer of gene 4, either deletion derivatives of the virulent Ti plasmid where the genes for hormone synthesis are inactivated or the "disarmed" Ti plasmid pGV 3850 into which gene 4 is inserted.

Our current investigations utilized the method of gene transfer which was elaborated earlier by some of the authors. It consists in the use of Ti plasmid pGV5 as the vector and the use of Ri plasmid as assistant. It gave the possibility to regenerate transgenic plants, which were constructed in the laboratory of Piruzyan and Andrianov [1]. As a result, the gene for synthesis of a cytokinin of the trans-ribosyl zeatin type was inserted into T-DNA of tobacco plants (*Nicotiana tabacum* cv. Samsun). Transgenic tobacco plants carrying gene 4 and rooting normally could be achieved by the use of this vector system.

## MATERIALS AND METHODS

*Escherichia coli* strains were cultivated on the medium of Luria and *Agrobacterium* strains on the YEB medium. Plant material was cultivated on solidified Murashige and Skoog (MS) [2] medium. Growth substances were omitted when not required. The medium of Gamborg [3]     was also used. *E. coli* strain K802, plasmids pGCO319, pRK2013 and *Agrobacterium* strain LBA9402 (pRi1855) were used throughout the experiments. Gene manipulations, consisting of isolation of plasmids, cleavage by restriction endonucleases, agarose gel, electrophoresis, and elution of specific DNA fragments. DNA ligation, *E. coli* transformation, and Southern blotting were performed according to the methods described elsewhere. The plant DNA was isolated according to the method described previously. The conjugation between *E. coli* and *A. rhizogenes* was performed on nitrocellulose filters.

Fresh tobacco leaves (cv. Samsun) of *in vitro* grown plants were used. *Agrobacterium* transformation was performed by the leaf disc method. Leaf discs were placed on MS medium with benzyladenine, kanamycin and claphoran for shoot induction. Shoots were rooted on Gamborg s medium with the same antibiotics. Rooted regenerants were assayed for neomycin phosphotransferase II (NPT II) activity.

The *E. coli* recombinant plasmid pGV0319, which carries whole T-DNA of the nopaline type Ti plasmid was used for the construction of a vector which transfers only gene 4 to the plant genome. This gene was cloned into pGK5, vector plasmid for binary *Agrobacterium* vectors, which was constructed in the laboratory of Piruzyan and

Andrianov [1]. The pGV0319 was cleaved by Barn HI and Hind III and the 2 kb fragment, which carries gene 4, was eluted from the electrophoresis gel. The pGK5 was cleaved by Eco RI and the previously isolated Bam HI - Hind III fragment was blunt and ligated into the vector. The cloned gene 4 remained under the control of its own promoter.

The newly constructed pGK5-5n4 plasmid was introduced into *A. rhizogenes* LBA9402, which carries a Ri plasmid by three strain conjugation. Exconjugants were selected on (YEB) medium with rifampicin, KM and carbenicilin. The presence of Bam HI - Hind III fragment was demonstrated by southern blotting, using the fragment as a probe. The fragment was detected in 4 out of 6 clones tested and one of them was used for tobacco transformation by the leaf disc method. Shoots resistant to kanamycin were found after 3-4 weeks.

Regenerated tobacco plants carrying gene 4 (as demonstrated by southern blotting) were cultivated in erlenmeyer flasks on solidified MS without growth substances. Plants were sectioned and one part of the material was used for clonal propagation and the other part for establishing callus cultures. The MS medium with IAA and BA was used for establishment and maintenance of callus cultures. Callus cultures of untransformed controls were established in parallel with transformants.

The levels of auxin and cytokinins in leaves and roots from sterile growing plants were estimated by following methods: the activity of the phytohormones was determined by biotests after purification of the methanolic extract of fresh material by ether or ethyl acetate and paper chromatography in the system isopropanol-ammonia-water 10:1:1. For auxin estimation segments of wheat coleoptile of cv. Albidum 43 were used; for cytokinin estimation the *Amaranthus* test.

The extraction and determination of IAA oxidase was carried out as described earlier. The composition of the test solution was as follows: 2ml 0.02M phosphate buffer pH 6.1, 1 ml. $MnCl_2$, 1 ml 1 mM 2,4-dichlorophenol, 2 ml 1 mM IAA and 4 ml enzyme extract. The enzyme activity was expressed in pg IAA oxidized per mg of protein in 1 min. Chlorogenic acid was isolated by two-dimensional chromatography using n-butanol acetic acid - water 4: 1: 1 and 15% acetic acid as solvents. The spot corresponding with the RF of a standard sample of chlorogenic acid was eluted and the absorbance of the eluate was measured at 280 nm by a spectrophotometer.

In order to obtain transgenic plants transformed by the T4 gene for cytokinin synthesis originating from the Ti plasmid of *Agrobacterium tumefaciens,* we applied the original vector system as described. In the binary vector system the Ri plasmid of wild type pRi 1855 was used as the helper plasmid. The utilization of such a plasmid brought some advantages in the construction of transgenic plants.

Tobacco (*Nicotiana tabacum* cv. Samsun) leaf discs were infected by a strain of *Agrobacterium* carrying the vector plasmid. After incubation, the resulting shoots were grown on a selective medium with kanamycin (KM). After three weeks of incubation poorly rooting dwarf plants were obtained. From ten plants tested by the NPT II assay, four gave a positive answer. Attempts to obtain morphologically normal regenerants on the MS medium or on Gamborg's medium, both containing 2,4-D, IAA, zeatin and other supplements, were unsuccessful. Therefore leaves of the original dwarf plants were used to get appropriate regenerants. From each of four dwarf plants giving a positive response in the NPT II assay, one leaf was detached and transformed on MS medium with BA, KM and CF. Shoots appeared after three weeks of incubation and they were transferred on Gamborg's medium without hormones containing KM, CF and 2% sucrose. Three shoots rooted without morphological deviations. Six regenerants from thirty were checked again by the NPT II test and showed the activity of neomycin phosphotransferase. DNA from one of the transgenic plants a positive result in the NPT II test was isolated and analyzed by Southern blotting. The plasmid pGKS, containing Bam HI-Hind III fragment, was used as a probe. Plant DNA was cleaved by Pst 1 and Barn HI-Eco RI restriction endonucleases. Using the enzyme Pst I characteristic fragments of 850 bp size were cut out.

In the other analysis restriction, endonucleases Barn HI and Eco RI were applied. By Southern hybridization the presence of gene 4 in T-DNA of the transgenic plants was confirmed. By this procedure phenotypically normal transgenic tobacco plants cv. Samsun transformed with the cytokinin synthesis gene in the T-DNA of transgenic plants were identified. In agreement with the reports the expression of gene 4 interferes with normal rooting of the transgenic plants. However, in the last experiments, phenotypically normal transgenic plants were obtained by the application of a vector system, using the Ri plasmid as helper.

Transformed tobacco plants were grown under sterile conditions on a solidified MS medium without hormones in a phytotron. Non-infected plants used as control were grown in the same aseptic conditions.

In order to prove by the physiological approach, the transformation of the plants and cuttings were excised and a part of them was used for derivation of callus tissue cultures. Callus formation was induced on an MS medium containing 3 pg IAA and 0.6 pg BA ml (MS+). The cultures were thereafter divided into two variants: the first variant was on a complete medium containing IAA and BA and the second variant on a selective medium without hormones (MS-). The callus cultures were cultivated in a 30-40 day sub-cultivation cycle either in light or in darkness. These paper only experiments with tissue cultures growing in the dark are described. As expected, the callus cultures derived from T4 plants grew well on the selective medium. In 5-6 subcultivations the tissue fresh mass reached an average value 4.5 ± 0.9 g. The callus derived from control plants did not grow on MS-medium (without hormones). Sometimes even tissues died in the first cultivation cycle. The Ti callus tissue grew well also on MS+ medium, i.e. 5.3 ± 1.4 g in one series. The calli derived from control plants also grew well on a full MS-medium forming a mass of 4.5 ± 0.8 g on average.

The content of endogenous cytokinins was determined in leaves and roots (Table **5**). In leaves of T4 plants the cytokinin content was five times higher than in the control. However, in roots of T4 plants, the level of cytokinins was much lower than in the control.

The results of GLC measurement showed a lower ABA content in the leaves of T4 plants than in leaves of the control. It could be assumed that the ABA level is to some extent influenced by the expression of gene 4. The ratio of auxin to cytokinins is of great importance for plant growth. Therefore, the level of endogenous auxin in leaves of transgenic and non-infected tobacco plants was determined.

Auxin levels increased ten times in the transgenic plants. The increase of the auxin content may be indirectly connected with the gene 4 expression, through the assumed influence of cytokinin on auxin synthesis. A second possibility is that not only was gene 4 transmitted from the vector plasmid but also T -DNA of pRi 1855 was transferred into the tobacco genome. This problem has not yet been carefully studied. As is known, the frequency of a common independent transfer of two different T -DNAs during infection by *Agrobacterium* is high. It was recently demonstrated that plant cells transformed by the Ri plasmid of *Agrobacterium rhizogenes* are more sensitive to auxin. It is possible that the transfer of the Ri T-DNA affect a compensation for the luxurious level of cytokinins induced by the T4 gene. As a result, phenotypically normal transgenic plants were formed. This situation has to be more closely studied.

The transformation of the plants is linked with a decrease of the oxidase activity both in leaves and in roots. The auxin protector chlorogenic acid was identified in the extracts and shown to be present at a higher level in extracts of T4 plant leaves (0.44 ± 0.02 pg/ g FW) in comparison with the control (0.12 ± 0.01 pg/g FW).

Our investigation demonstrated changes in the hormone levels in tobacco plants after insertion of the T4 gene into the plant genome. Observed changes in the transgenic plants are of a complex nature. Simultaneously with an enhancement of cytokinin and auxin levels, there was a decrease in the content of ABA. The auxin oxidase activity in the transgenic plants was lowered, while at the same time the content of an auxin protector, chlorogenic acid, increased in the transformed plants.

# REFERENCES

[1]    Piruzyan EI, Andrianov VM. Analysis of replicon of Ti-plasmid of nopaline type *Agrobacterium tumefaciens* C58 and the possibility of its use for transfer of genes into plants (In Russ.). Genetika 1986; 22: 2674-2683.
[2]    Murashige T, Skoog F. A revised medium for rapid growth and bioassays with tobacco tissue culture. Physiol Plant 1962; 14: 473-497.
[3]    Gamborg OL, Eveleigh DE. Culture methods and detection of glucanases in suspension cultures of wheat and barley. Can J Biochem 1968; 46: 417-421.

# CHAPTER 12

# Phytohormones, Genome and Properties

**Abstract:** This chapter deals with the different properties of the growing plants. Starting point is photosynthesis. Products of this process are connected with different forms of growth processes, which are under the genetic control. Dwarf and semi dwarf plants could absorb different amount of mineral nutrients from the soil. Genetically modified plants have different amount of plant hormones and inhibitors.

## INTRODUCTION

The process of plant growth presents some of the most attractive problems of physiology and is one of the advanced areas in which it can be said that we have made a crucial breakthrough on the basis of achievements of plant hormonology [1, 2]. In this paper, which is an extract from the lecture dedicated to the memory of Kholodny, we tried to concentrate on the hormonal aspects of development, including ideas of gene control of this phenomenon [3, 4].

There are large gaps in our understanding of some of the most important aspects of the chain "hormones-genome-properties" such as:

- hormonal control of gene expression,
- genetic control of hormone biosynthesis,
- hormonal changes in mutants, transformants, and haploids under changes in development.

It is clear that the main properties of growth and development should involve a certain control mechanism. As a result of the intensive studies of many years, it is now clear that hormones play a vital role in the control of growth, not only within the plant as a whole, but apparently also within individual organs [5-9].

Now it is known that there are at least three major classes of growth promoting hormones, namely auxins, gibberellins, and cytokinins. In addition, other classes of plant hormones exist, particularly the "growth inhibitors" such as abscisic acid (ABA), but also ethylene which is apparently involved in many growth phenomena [9-11].

The fact that developmental processes are basically hormonal and gene-controlled is self-evident, since there are genetic and hormonal variations which affect almost every aspect of development ranging from the external morphology and internal anatomy to physiological characters such as growth rate, blossom time and of the dormancy period duration [12-17]. Growth and development are such orderly processes that must require the right genes to be expressed in the right cells at the right time. That is to say, growth and development are essential processes involving selective gene expression, and the concept that these processes involve the activity of specific groups of genes which in turn control the synthesis of some enzymes including those regulating hormonal and inhibitory biosynthesis characteristic of some specialized cells, is called the variable gene expression theory of growth and differentiation [18, 19]. Hormonal substances and their phenolic modificators can regulate not only growth and development processes but also photosynthesis [20, 21]. The visible manifestations of growth and differentiation include variations in the development of the cell wall and of certain cytoplasmic organelles such as plastids. It is clear that differentiation must also extend to certain aspects of metabolism when one keeps in mind that some tissues are specially adapted to particular functions, such as photomorphogenesis, photosynthesis, secretion and storage of reserve materials [22-26].

Now we would like to describe the hormonal status of some mutants and transformants in order to clarify the concept "genome hormones properties" of various plant forms and in the connection between growth and photosynthesis [27-29]. It is no doubt, that the precursors of plant growth regulators are tightly connected with the primary products of photosynthesis [30, 31]. Thus, photosynthesis plays a role of source for phenolics, auxins, ABA and gibberellins (See the scheme below).

**Narcin Palavan-Unsal (Ed)**

```
┌─────────────────────────────────────────────────────────────────────┐
│  PRIMARY PRODUCTS OF PHOTOSYNTHESIS AND GROWTH REGULATORS             │
│                                                                       │
│        Schikimate                       Acetate-mevalonate            │
│            │                                   │                      │
│            ↓                                   ↓                      │
│        Chorizmate                   Geranylgeraniol phosphate         │
│          ↙   ↘                          ↙            ↘                │
│  Tryptophan  Phenylalanine Tyrosine  ┌──────────┐  ┌────────────┐    │
│      │            │                  │Abscisic acid│ │Gibberellins│   │
│      ↓            ↓                  └──────────┘  └────────────┘    │
│  ┌──────────┬──────────────────┐                                     │
│  │Indolic auxins│ Phenolic inhibitors│                               │
│  └──────────┴──────────────────┘                                     │
└─────────────────────────────────────────────────────────────────────┘
```

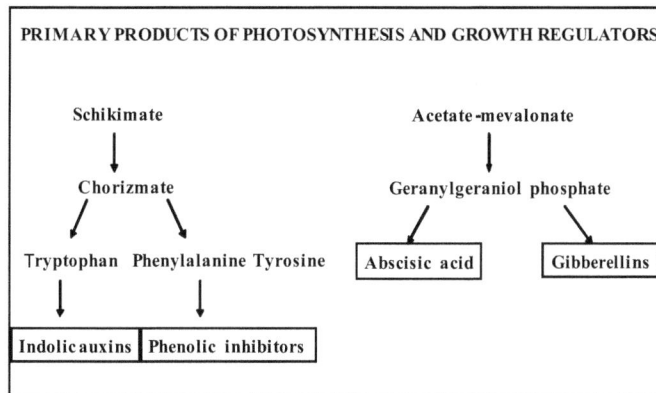

This connection is not always covered by one step and is not obviously specific. There are some evidences of effects of IAA on the phenolic formation as well as back effects. Feedback effect of cinnamic acid on phenolic biosynthesis exists also. Cinnamic acids inhibit intracellular PAL activity. Whether feedback inhibition of phenylalanine-ammonia-lyase (PAL lyase) by cinnamic acids has physiological significance in buckwheat, is, however, questionable, as free cinnamic acids do not accumulate in the tissue and may thus be available in sufficiently high concentrations to cause inhibition of PAL [32]. The end products of cinnamate metabolism in buckwheat, flavonol glycoside, rutin. and despite chlorogenic acid had little, if any, effect on intracellular PAL-activity. The crucial question is, however, if these compounds penetrate *in vivo* to the intracellular site of PAL, and thus any definite statement as to the absence of *in vivo* feedback inhibition of PAL by these products cannot make (see, however, statement on feedback repression below).

As phenolic compounds have been shown to interfere with indole biosynthesis, the opposite situation, i.e. inhibition of phenol synthesis by indole compounds was considered to occur possibly at the level of PAL. The precursors of L-tryptophan, anthranilic and indolepyruvic acids, had a peculiar effect on $^3$HOH formation from radioactive phenyl-$C^{14}$ alanine firstly by repressing the formation and then, after a lag phase of about 90 min, allowing it to proceed at the rate of the control. The common precursor of the three aromatic amino acids, shikimic acid, showed, however, no effect, while the aromatic amino acids, tyrosine and tryptophan, again caused inhibition. As in this type of experiment the effect of an exogenously added compound on the rate of equilibration of the similarly exogenously supplied substrate. phenylalanine, with the endogenous substrate pool of PAL is unknown and cannot be determined, the interpretation of the results can necessarily not be unequivocal. The result obtained with tryptophan prompted experiments with indoleacetic acid and other auxins, which are described below.

Indoleacetic acid (IAA) was found to inhibit the formation of HOH from labeled phenylalanine with a potency similar to that of the cinnamic acids, while the auxin-analogues, ct-NAA and 2,4-D, were less active. Surprisingly, the inactive auxin-analogue, β-NAA showed reproducible inhibitory activity considerably higher in the intact cell assay than the active auxin-analogue α-NAA, indicates that the effect is not completely specific for auxins. All compounds tested were found to be competitive inhibitors of buckwheat PAL extracted from an acetone powder of illuminated hypocotyls.

IAA and β-NAA were the most potent inhibitors, which is in agreement with the data obtained from the intact cell assays. Inhibition of PAL-activity *in vivo* should result in the reduced accumulation of metabolites derived from cinnamic acid. Light-induced anthocyanin formation was used as an indicator of the *in vivo* production of cinnamic acid, as the anthocyanin content of a tissue can be determined by rapid and simple procedures. All compounds tested reduced the production of anthocyanins in isolated hypocotyls incubated in the light, but in these experiments α-NAA showed a nearly tenfold higher inhibitory activity as compared to β-NAA. This result clearly indicates that the effect of auxins on anthocyanin biosynthesis is rather complex and involves more than one site of action. That inhibition of intracellular PAL by high concentrations of exogenous auxin may, nevertheless, be partially involved in the inhibition of anthocyanin synthesis IS made likely by the fact that the IAA-mediated inhibition can fully or partially be overcome by the application of phenylalanine or cinnamic acid. It is necessary to mention that growth depression sometimes is accompanied by phenolics accumulation. This situation could be observed, for example, when pea plants were exposed to high-intensity light (xenon arc lamp). The suppression of growth was accompanied by an increase in the quantities of quercetin derivatives, of which quercetin-3 glucosyl-p-coumarate (QGC) was present in the greatest amount (75%). Exogenous application of p-coumaric acid (PCA) and QGC to pea stem segments depresses their growth [30, 33]. However, in pea leaves the rate of photosynthesis goes up until the light intensity of 200.000 erg/em sec, and then remains at the same level Table **1**).

**Table 1:** Interference of various compounds with intracellular formation of phenylalanine by segments of buckwheat hypocotyls under red light illumination.

| Compounds | Effect of $^3$HOH formation from L-(3-$^3$H)-phenylalanine |
|---|---|
| Cinnamic acid | Inhibition: $I_{50}$ 0.05 mM |
| p-Coumaric acid | 0.1 mM (50% inhibition) |
| Caffeic acid | Inhibition: 0.1mM |
| Chlorogenic acid | 10-20% inhibition with 1mM |
| Rutin | No effect with 1mM |
| Shikimic acid | 50% inhibition with 1mM |
| Tyrosine | In the presence of 1mM, rate of $^3$HOH |
| Anthranilic acid | Formation is identical with that of control |
| Indolepyruvic acid | After 90 min. lag-phase same as for anthranilic acid |
| Indole | 50% inhibition with 1mM |
| L-tryptophan | 80% inhibition with 1mM |
| D L-p-fluorophenyl-alanine | 100% inhibition with 0.5mM |

The comparison of data obtained on the photosynthesis and growth of plant evidences that at light- saturating intensities the reduction in the total plant productivity is mainly due to the inhibition of the leaf area growth and, to a lesser extent, to a lower rate of photosynthesis.

It might be thought that the accumulation of photosynthetic products (photosynthates) during inhibited growth occurs in the same way as in the conditions of vigorous stem extension. However, the use of these products, including phenolic compounds, for cell lignification processes and for building up a cell skeleton during elongation is greatly decreased. Phenol derivatives and p-coumaric acid which is one of the leading products of this type acquire a new function-growth inhibitory because they are not fully used for lignification of elongated cells. This phenomenon can be demonstrated by the following scheme.

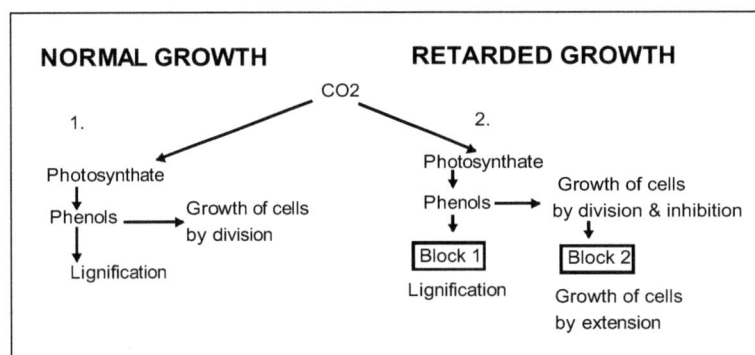

Thus, the juvenile period in plant ontogeny is a heterogeneous stage of development which involves repeated blocking of growth processes with the result that one type of growth is switched over to another. In this situation, some phenolics could playa role of markers or factors regulating the growth.

Now we decided to compare behavior of dwarf mutants in the light of various intensity. Fourteen day-old pea plants cv. Torstag (tall) and the mutant's semi dwarfK-29 and dwarfK-202 were analyzed to reveal the QGC level. Dark grown plants had the tallest stems, their leaf plates were reduced. At light (50, 200, 400 W. m") the stem growth of all pea forms was suppressed. Their leaf plate became larger, but in light of high intensity its area decreased. Inhibition of stem growth is enhanced with increasing light intensity. The dwarf forms with stems were more sensitive to light even of low intensities. In light of high intensity, the stem of the tall variety was shortened and "light" dwarf plants were formed. Inhibition of stem growth as result of mutation or light action is correlated with the thickening of leaf plate, because the number of layers of mesophyll cells is increased (Table **2**).

The change of area and anatomical structure of the leaf can be the reason for an increase of photosynthetic activity. The content of gibberellins was decreased in all plant forms with suppressed stem growth.

**Table 2:** Anatomical characteristics of leaves of 40-days old original and mutant pea plants.

| Thickness | Light intensity, W. $M^{-2}$ | | |
|---|---|---|---|
| | 50 | 200 | 400 |
| **Torstag** Leaf | $88.7 \pm 1.00$ | $116.9 \pm 0.09$ | $168.5 \pm 1.05$ |
| Palisade mesophyll | $30.2 \pm 1.00$ | $40.1 \pm 1.03$ | $57.0 \pm 1.20$ |
| Spongy mesophyll | $41.9 \pm 2.30$ | $57.6 \pm 1.63$ | $87.4 \pm 2.71$ |
| **Mutant 202** | | | |
| Leaf | $190.5 \pm 3.21$ | $270.0 \pm 4.49$ | $303.5 \pm 4.10$ |
| Palisade mesophyll | $82.6 \pm 1.72$ | $106.7 \pm 1.99$ | $120.4 \pm 2.30$ |
| Spongy mesophyll | $91.3 \pm 2.27$ | $140.7 \pm 3.90$ | $161.2 \pm 1.40$ |

However, depression of stem growth induced by mutation or light action is correlated with an increase of QGC accumulation up to 20-30 mg.g-$^{1}$ of dry matter. This is probably caused by an increase of photosynthetic activity in thickened leaves of the dwarfs. For solving this question we used a xanthomutant of pea.

This mutant is characterized by a block in chloroplastogenesis. Plastids of this mutant differ from normal ones by smaller size, more spherical form, and absence of lamellar structure. Therefore photosynthesis was completely suppressed. Eight day old mutants did not differ in their size from original ones.

It is possible to conclude that all morphogenetical reactions except chloroplastogenesis are developed similarly in both forms of plants. The block of chloroplastogenesis in the mutant caused inhibition of chlorophyll synthesis and photosynthesis. Carotenoids are not synthesized in etiolated plants, but in light their content in xanthomutants was about 12% in comparison to green ones. The content of phenolics calculated on fresh and dry matter shows similar levels in xantho and green plants (Table **3**).

**Table 3:** Content of phenolics in pea xanthomutant.

| Type and Condition | Content of phenolics (mg/l) | | | |
|---|---|---|---|---|
| Normal at light | 5.0 | 36.3 | 0.04 | 0.30 |
| Normal in darkness | 1.3 | 17.1 | 0.30 | 0.40 |
| Mutant at light | 5.7 | 36.9 | 0.04 | 0.20 |
| Mutant at darkness | 1.2 | 17.5 | 0.02 | 0.30 |

Dwarf mutations and light of high intensity increased the QGC-level. In order to find out whether there are differences in metabolism of the QGC precursor PCA in the different types of pea plants this was studied in the variety Torstag at light 200 W.m$^{-2}$ (tall) or light 400 W.m-$^{1}$ (light dwarf) and in mutant dwarf plants K-202 at light 200.

The mutant dwarf and the "light" dwarf plants possess similar catabolism of PCA. Considerable incorporation of radioactivity into the methanol-unsoluble fraction which includes generally cell walls was shown. Maximal C-incorporation was observed in tall Torstag plants (28.7% of total incorporation) while it was less prominent in light (20.9%) and mutant (18.3%,) dwarf plants. C-incorporation into low-molecular methanol-soluble metabolites was the lowest in the tall plants (Table **4**). Methanol extracts from light and mutant dwarf plants were enriched with radioactive metabolites, first of all, with glucose ester of PCA.

**Table 4:** Radioactivity of methanol-soluble metabolites of PCA from pea.

| Variety, light intensity, W.m$^{-2}$ | Radioactivity, % of methanol extracts | | | |
|---|---|---|---|---|
| | Free PCA | Glucose ester PCA | QGC | Non-identified metabolite |
| Torstag, 200 | 58.48 ± 2.9 | 24.77 ± 1.2 | 4.15 ± 0.2 | 12.54 ± 0.6 |
| Torstag, 400 | 42.97 ± 2.1 | 37.33 ± 0.6 | 6.89 ± 0.3 | 12.81 ± 0.6 |
| K-202, 200 | 37.96 ± 1.9 | 40.50 ± 2.0 | 6.61 ± 0.3 | 14.93 ± 0.7 |

What is the mode of action of PCA, quercetin and QGC on growth? QGC inhibited the growth of pea stem sections only at 8.000 mgl$^{-1}$ while the growth of wheat coleoptile sections - at 4.000 mg. l$^{-1}$. The concentration of endogenous conjugated flavonoids is very high in pea tissues. Quercetrin (quercetin-glucoside) was not able to inhibit growth of wheat coleoptile sections even at 220 mg. l$^{-1}$ concentration (semi saturated solution). PCA started to depress the growth at 175 mg/l. QGC, which consists of inert quercetin glucoside and of active PCA was able to depress the growth of wheat coleoptile sections at a concentration 20 times higher than that of PCA. However, PCA is capable to activate cell division in tobacco suspension cultures.

Pea plants contain large amounts of QGC. The content of free PCA, other hydroxycinnamic acids, and their glucose esters is much lower. However, concentration of QGC changes in wide limits and depends on the variety of pea plants. Dwarf forms induced by mutation or light action are accompanied by a high QGC accumulation.

When stem growth was blocked, striking changes were observed in the leaves. Leaf thickness increased as a result of an increased number of parenchyma and mesophyll cell layers. It is accompanied by an increase of photosynthesis activity. Is it possibly a reason of QGC accumulation?

The experiments with the xanthomutant of pea have shown that formation of phenolic compounds is not dependent on photosynthesis and structural organization of chloroplasts. Probably all of these processes proceed under direct control by light through the activation of PAL and 4-cinnamic acid hydroxylase. But accumulation of QGC in peas is accompanied by depressed stem growth. The metabolism of 2-$^{14}$C-trans-PCA in "light" and mutant dwarf plants is similar and distinguished from that in tall plants. Hydroxycinnamic acids are known to be lignin and flavonoid precursors. PCA incorporation in methanol-unsoluble material can be explained as constituent of cell was lignin.

Formation of conjugated forms of PCA (glucose ester and QGC) was higher in "light" and mutant dwarf plants. Such increase of glucose ester of PCA in comparison with free PCA was observed at blue light.

We suppose a close relation between the formations of glucose esters of PCA and stem growth depression. It is possible that the incorporation of hydroxycinnamic acid into cell walls of dwarf plants is inhibited. In cell wall formation, non utilized precursor accumulates as methanol-soluble, low-molecular conjugates which are able to affect cell elongation and division. The latter is especially interesting because depression of stem elongation is accompanied by activation of mesophyll cell division.

Thus, differences in QGC levels and in PCA metabolism in tall and dwarf pea plants, absence of direct connection with photosynthesis and participation in some growth processes show an important role of different phenolic conjugates in common metabolism of plants and in growth regulation.

Thus, dwarfism is tightly connected with the following systems "auxin-phenolics" and gibberellin, ABA However, the gene control of cytokinins is investigated on the other model [34].

In the last few years, transgenic plants with integrated bacterial gene 4 have been obtained by this procedure. It was shown that regenerated plants with transferred gene 4 are dwarfed and rooted with difficulty. The authors of these studies used for the transfer of gene 4, either deletion derivatives of the virulent Ti plasmid where the genes for hormone synthesis are inactivated, or the "disarmed" Ti plasmid pGV 3850 into which gene 4 is inserted.

Our current investigations utilized the method of gene transfer which was elaborated earlier by some of the authors. It consists the use of Ti plasmid pGV5 as the vector and the use of Ri plasmid as an assistant. It gave the possibility to regenerate transgenic plants, which were constructed in the laboratory of Piruzyan. As a result the gene for synthesis of a cytokinin of the trans-ribosyl zeatin type was inserted into T-DNA of tobacco plants *(Nicotiana tabacum* cv. Samsun). Transgenic tobacco plants carrying gene 4 and rooting normally could be obtained by the use of this vector system.

The newly constructed pGK5-5n4 plasmid was introduced into *A. rhizogenes* LBA9402, which carries to Ri plasmid, by three strain conjugation. Exconjugants were selected on (YEB) medium with rifampicin 50 μg m $l^{-1}$, KM 12.5 μg m$l^{-1}$ and carbenicilin 100 μg. The presence of Bam HI-Hind III fragment was demonstrated by Southern blotting, using the fragment as a probe. The fragment was detected in 4 out of 6 clones tested and one of them was used for tobacco transformation by the leaf disc method. Shoots resistant to kanamycin were found after 3-4 weeks.

Regenerated tobacco plants carrying gene 4 (as demonstrated by Southern blotting), were cultivated in Erlenmeyer flasks on solidified MS without growth substances. Plants were sectioned and one part of the material was used for clonal propagation and the other part for establishing callus cultures. The MS medium with IAA and BA was used for establishment and maintenance of callus cultures. Callus cultures of untransformed controls were established in parallel with transformants.

The levels of auxin and cytokinins in leaves and roots from sterile growing plants were estimated by the following methods: the activity of the phytohormones was determined by biotests after purification of the methanolic extract of fresh material by ether of ethylacetate and paper chromatography in the system isopropanol-ammonia-water (10:1:1). For auxin estimation segments of wheat coleoptile of cv. Albidum 43 were used; for cytokinin estimation, the *Amaranthus* test was performed. The ABA content was measured by the gas-liquid chromatography (GLC) method.

The extraction and determination of IAA oxidase was carried out as described earlier Chlorogenic acid was isolated by two-dimensional chromatography using n-butanolacetic acid: water, 4: 1 and 15% acetic acid as solvents. The spot corresponding to the Rf of a standard sample of chlorogenic acid was eluted and the absorbance of the eluate was measured at 280 nm by a spectrophotometer.

Now, the formation of cytokinins is under the distinct gene control. It is important to mention, that in the system phenolics also play their role of modifiers of the hormonal activity. The previous genetic modifications concern the growth or hormonal content. Now we consider the photosynthetic changes during the genome mutation [35, 36].

In order to obtain transgenic plants transformed by the T4 gene for cytokinin synthesis originating from the Ti plasmid of *Agrobacterium tumefaciens,* we applied the original vector system as described. In the binary vector system the Ri plasmid of wild type pRi 1855 was used as the helper plasmid. The utilization of such a plasmid brought some advantages in the construction of transgenic plants.

Tobacco *(Nicotiana tabacum* cv. Samsun) leaf discs were infected by a strain of *Agrobacterium* carrying the vector plasmid. After incubation the resulting shoots were grown on a selective medium with kanamycin (KM). After three weeks of incubation, poorly rooting dwarf plants were obtained. From ten plants tested by the NPT II assay, four gave a positive answer. Attempts to obtain morphologically normal regenerants on the MS medium or on Gamborg's medium, containing 2.4-D, 1AA, zeatin and other supplements, were unsuccessful. Therefore, leaves of the original dwarf plants were used to get appropriate regenerants. From each of four dwarf plants giving a positive response in the NPT assay, one leaf was detached and transformed on MS. Three shoots rooted without morphological deviations. Six regenerants were checked again by the NPT II test and showed the activity of neomycin phosphotransferase. DNA from one of the transgenic plants positive in the NPT II test, was isolated and analyzed by Southern blotting. The plasmid pGKs, containing Barn HI-Hind III fragment, was used as a probe. Plant DNA was cleaved by Pst 1 and Barn HlEco RI restriction endonucleases. Using the enzyme Pst I characteristic fragments of 850 by size were cut out.

In the other analysis restriction, endonucleases Barn HI and Eco Rl were applied. By Southern hybridization the presence of gene 4 in T -DNA of the transgenic plants was confirmed. By this procedure, phenotypically normal transgenic tobacco plants cv. Samsun transformed with the cytokinin synthesis gene in the T-DNA of transgenic plants were identified. In agreement with the reports, the expression of gene 4 interferes with normal rooting of the transgenic plants. However, in the last experiments, phenotypically normal trans-genic plants were obtained by the application of a vector system, using the Ri plasmid as helper.

Transformed tobacco plants were grown under sterile conditions on a solidified MS medium without hormones in a phytotron. Non-infected plants used as control were grown in the same aseptic conditions.

In order to prove by the physiological approach of the transformation of the plants, cutting was excised and a part of them used for derivation of callus tissue cultures. Callus formation was induced on a MS medium containing 3 mkg IAA and 0.6 mkg BA (MS+). The cultures were thereafter divided into two variants: first variant on a complete medium containing lAA and BA and a second variant on a selective medium without hormones (MS-). The callus cultures were cultivated in a 30-40 day sub-cultivation cycle either in light or in darkness. In this paper only experiments with tissue cultures growing in the dark are described. As expected, the callus cultures derived from T4 plants grew well on the selective medium. In 5-6 sub-cultivations, the tissue fresh mass reached an average value 4.5 ± 0.9 g. The callus derived from control plants did not grow on MS- medium (without hormones). Sometimes even tissues died in the first cultivation cycle. The Ti callus tissue grew well also on MS+ medium, i.e. 5.3 ± 1.4 g in one series. The callus derived from control plants also grew well on a full MS[4], medium forming a mass of 4.5 ± 0.8 g on average. The content of endogenous cytokinins was determined in leaves and roots (Table **5**). In leaves of T4 plants the cytokinin content was five times higher than in the control. However, in roots of T4 plants the level of cytokinins was much lower than in the control.

**Table 5:** The activity of endogenous cytokinins in normal and transgenic tobacco plants.

| Plant | Organ | Cytokinins (mkg/l) |
|---|---|---|
| Control | Leaves | 0.16 ± 0.01 |
| | Roots | 0.14 ± 0.01 |
| T4 | Leaves | 0.86 ± 0.03 |
| | Roots | 0.02 ± 0.004 |

The results of GLC measurements showed a lower ABA content in the leaves of transformed plants than in leaves of the control. It could be assumed that the ABA level is to some extent influenced by the expression of gene 4 (Table **6**).

**Table 6:** Abscisic acid content in normal and transgenic tobacco leaves.

| Plant | ABA (mkg/l) |
|---|---|
| Control | 75.7 ± 4.5 |
| T4 | 45.5 ± 1.8 |

The ratio of auxin to cytokinins is of great importance for plant growth. Therefore, the level of endogenous auxin in leaves of transgenic and non-infected tobacco plants was determined.

Auxin levels increased ten times in the transgenic plants. The increase of the auxin content may be indirectly connected with the gene 4 expression, through the assumed influence of cytokinin on auxin synthesis. A second possibility is that not only was gene 4 transmitted from the vector plasmid but also Ti-DNA of pRi 1855 was transferred into tobacco genome. This problem has not yet been carefully studied. As it is known, the frequency of a common independent transfer of two different Ti-DNAs during infection by *Agrobacterium* is high. It was recently demonstrated that plant cells transformed by the Ri plasmid of *Agrobacterium rhizogenes* are more sensitive to auxin. It is possible that the transfer of the Ri Ti-DNA affects a compensation for the luxurious level of cytokinins induced by the T4 gene. As a result, phenotypically normal transgenic plants were formed. This situation has to be more closely studied.

The transformation of the plants is linked with a decrease of the oxidase activity both in leaves and in roots. The auxin protector chlorogenic acid was identified in the extracts and shown to be present at a higher level in extracts of

T4 plant leaves (0.44 ± 0.02 picogram per L) in comparison with the control (0.12 ± 0.01 picogram per L) (Tables **7** and **8**).

**Table 7:** Auxin oxidase activity in normal and transgenic tobacco plants.

| Plant | Auxin (mkg/l) |
|---|---|
| Control | 0.02 ± 0.002 |
| T4 | 0.21 ± 0.009 |

**Tables 8:** Auxin oxidase activity in normal and transgenic tobacco plants.

| Plant | Organ | Oxidase Activity (IAA destroyed Mill mg$^{-1}$ protein) |
|---|---|---|
| Control | Leaves | 34.4 ± 2.2 |
|  | Roots | 62.1 ± 15.2 |
| T4 | Leaves | 24.1 ± 3.9 |
|  | Roots | 32.2 ± 8.1 |

Our investigation demonstrated changes in the hormone levels in tobacco plants after insertion of the T4 gene into plant genome. Observed changes in the transgenic plants are of a complex nature. Simultaneously with an enhancement of cytokinin and auxin levels, there was a decrease in the content of ABA. The auxin oxidase activity in the transgenic plants was lowered, while at the same time the content of an auxin protector, chlorogenic acid, increased in the transformed plants.

The problem of productivity of green plants is based on the accumulation of biological mass and on the crop formation. Both processes are the result of cooperation of growth and photosynthesis. Exclusion or depression of one of these processes inhibits the normal productivity, and on the other hand allows a better understanding of the functions of the coupled process. Thus, the inhibition of stem growth by mutagenes resulting in dwarf forms of pea and cotton permits us to observe the stability of net photosynthesis, which does not change remarkably in the leaves of dwarf mutants.

**Table 9:** Chloroplast pigments in cotton and mutants.

| Forms | Chlorophyll (a and b) | | Carotenoids | |
|---|---|---|---|---|
|  | Mg/g fr.mass | Mg/dry mass | Mg/g fr.mass | Mg/g dry mass |
| **Cotton** | | | | |
| Green | 7.60 | | 2.80 | |
| Etiolated | 0.02 | | 0.01 | |
| Albino (grown in light | 0.47 | | 1.30 | |
| **Pea** | | | | |
| Green | 1.318 | 9.400 | 0.356 | 2.500 |
| Etiolated | 0.002 | 0.026 | 0.002 | 0.026 |
| Albino (grown in light | 0.019 | 0.122 | 0.044 | 0.290 |
| Albino (grown in dark) | 0.02 | 0.026 | 0.005 | 0.070 |

Now we have just the opposite task to investigate albino mutants with depressed photosynthesis. The growth of green and albino seedlings was investigated during the first week and a similarity was found (Table 9).

The starting material for albino plants are two lines: L-73 and L-453 from the collection of *Gossypium hirsutum* L. of Tashkent University. The authors express their thanks to Musaev and Almatov for supplying these plants. In 1977 lines L-73 and L-453 were bred at Tashkent University and hybrid seeds F0 were obtained. The last were treated by 0.25% nitrosomethylurea and sown in 1978 as F1.In the population of M1 and M2 the splitting (3:1) at the absence of chlorophyll was observed. All plants of this family were self-pollinated and the seeds obtained were sown again in 1980 as F3. In our experiments we used seeds of F3.

Pea albino mutants were obtained in the same way using 0.025% solution of nitroso methylurea. As the albino mutants possess carotenoids, they were called xanth-mutants by the authors. Plants were grown under phytotronic conditions. Temperature 24°C, 70% air humidity, in white light. Luminescent lamps were used as a source of illumination. The age of experimental plants was 7 days for pea and 8-10 days for cotton.

The growth of cotton plants (10 d old) green and albino was similar (green - 11.9 cm, albino 10.9 cm). However, the photosynthesis of albino plants was not observed. Chloroplast pigments of cotton mutants were reduced. Although we could not observe the photosynthetic activity of albino mutants, the flavonoid content in these plants was similar to that in the green ones (Table **10**), and anthocyanins were diminished only twice. No dramatic difference was observed in the content of auxins (Table **11**) and ABA level in albino mutants was even higher than in green plants.

Thus, our observations on the flavonoids, anthocyanins and the hormonal level of albino mutants show that they probably have preserved the main photomorphogenetic reactions, typical of the green plants. However, the genesis of chloroplasts is depressed.

**Table 10:** Flavonoids and anthocyanins in cotton mutants (mg g$^{-1}$ of fresh mass).

| Forms | Flavanol glucoside (mg g$^{-1}$ of fresh mass) | Anthocyanin (mg g$^{-1}$ of fresh mass) |
|---|---|---|
| Green | 13.0 | 20.0 |
| Etiolated | 0.0 | 0.0 |
| Albino (grown in light) | 12.5 | 10.0 |

**Table 11:** Phytohormones in 10-days cotton mutants.

| Forms | ABA*** | Auxins* | Inhibitors** |
|---|---|---|---|
| Green | 155 | 65 | 1.0 |
| Albino | 125 | 60 | 2.5 |

* Biological activity of IAA Rf region (% of control)

** Biological activity of ABA Rf region (% of control)

*** GLC determination of ABA (Relative units according to pike square)

The growth of pea albino mutants (7 d old) was similar to the green ones. By cultivation in darkness they were also subjected to etiolation (Table **12**). Chlorophylls in etiolated and in light-grown albino plants are practically absent (Table **9**).

**Table 12:** Phytohormones in pea mutants.

| Forms | Auxin* | Inhibitors** |
|---|---|---|
| Green | 124 | 85 |
| Etiolated | 127 | 94 |
| Albino (grown in light) | 122 | 80 |
| Albino (grown in dark) | 130 | 108 |

* Biological activity of IAA Rf region (% of control)

** Biological activity of ABA Rf region (% of control)

Yellow (albino) pea forms are not able to photosynthesize. The activity of photosynthesis was determined by evaluation. Albino plants do not possess remarkable fluorescence, which could inform us about changes in plastid structure. The albino plants cultivated in light do not form any noticeable amount of chlorophyll, and they had a highly reduced level of carotenoids (Table **9**). It is necessary to mention that these values do not correlate with the level of phenolics.

The content of phenolics calculated on fresh and dry mass shows similar levels; the content of QGC and p-coumarate was equal in albino (xantho) and green plants. No distinct differences were observed in the level of auxins and in inhibitors (Table **12**). It is to be mentioned that etiolated plants do not contain any appreciable amount of natural inhibitors, only auxins are present in a significant concentration. These facts were in coincidence with results obtained with cotton albino mutants.

The xantho mutants and original green plants had similar characteristics in their reactions to the light. Xantho mutants preserved all morphogenetic reactions concerning the growth, although they lost all ability concerning chlorophyll pigment synthesis.

In albino mutants the level of auxin and growth-inhibitors was not altered remarkably. It is necessary to mention that the level of ABA in albino cotton plants was 2.5 times increased in respect to normal green plants. In pea plants, there was no depression of phenolics, only anthocyanins in cotton albino plants decreased twice. Thus, there is no correlation between the level of chloroplast pigments in albino plants of pea and cotton and the activity of phytohormones (auxin and ABA) and level of phenolics.

Considering the biosynthetic aspects of ABA and phenolics, these two groups of sub-stances are known to be synthesized in chloroplasts, According to our results, their level does not change remarkably under the depression of the genesis of chloroplasts, chlorophyll and carotenoid formation. Why? There are two hypothetical possibilities:

a.   either one additional biosynthetic center exists, or

b.   the reduced level of chloroplast pigments is enough for the formation of phenolic substances and ABA in protoplasts. Until now these questions have not been solved. Comparing the results of analyses of etiolated and albino plants another idea arises. Albino plants are sensitive to the light and show the following morphogenetic reactions:

   a.   depression of stem growth;

   b.   activation of cotyledon (cotton) or leaf (pea) development;

   c.   activation of anthocyanin and flavonoid synthesis;

   d.   phototropism.

All these reactions are known to be regulated by phytochrome system. It follows that all these reactions are not directly connected with photosynthesis and genesis of chloroplasts. Concluding our results, we propose the following scheme for the investigated processes involved in the productivity of plants.

| Genes responsible for genesis of chloroplasts | Non-active (albino plants) | genes | Genes responsible for growth | Genes responsible for photomorphogenesis |
|---|---|---|---|---|
| Mutagens (no photosynthesis) | Mutation | | Active process | Active process |

In our experiments mutagens have induced the mutation of genes responsible for the genesis of chloroplasts, nevertheless in the resulting albino plants, the gene blocks inducing growth, morphogenesis, synthesis of phytohormones, and phenolics were still acting.

## CONCLUSION

The concept of Cholodny of the system of hormones which regulate ontogeny is now developed in the way of gene control of the whole hormonal system as a primary regulating block [3, 4, 37-39].

It is important to mention that the gene controls hormonal signal on the one hand, and hormones and their analogues are able to control gene expression on the other [40-43]. There are many approaches to the search for the effect of genome on hormonal level and vice versa [44, 45].

It is also important to mention that one gene change induces various chains of growth regulators modifications which must be investigated carefully in the future.

## REFERENCES

[1]     Kholodny NG. Selected Works. Kiev, Acad. of Sci. Ukrainian SSR. Publ. House, 1956;

[2]     Chailakhyan M. Regulation of Higher Plant Blooming. Moscow 1978.

[3]     Gleba YY, Sytnik KM. Plant Cell Engineering. Kiev 1984.

[4]     Piruzyan ES. Principles of Plant Gene Engineering. Moscow, Nauka. 1988.

[5]     Kefeli VI, Turetskaya RK. Evolutionary mechanisms for formation of hormonal regtillation system in plants in connection with growth and morphogenesis. Problems of Evolutionary Plant Physiology. Leningrad. Leningrad. Otd. In-ta istorii Estestvoznaniya i Tekhniki Akad. Nauk SSSR, 1974; 105-106.

[6]     Syrtanova GA, Rakhimbaev IR, Kefeli VI. Tulip native growth inhibitor. Fiziologiya rastenii, 1975; 22: 165-169.

[7]     Solomina VF, Rakhimbae IR, Kefeli VI, Komizerke EI. Isolation and identification of native cytokinin-zeatin riboside from growing tulip bulbs. Dokl. Akad. Nauk USSR, 1976; 229: 1274-1276.

[8]     Sudeinaya SV. Effect of various methods of treatment with auxins and fungicide on rooting of remontant carnation cuttings. In: Plant Growth and Development Regulators. Moscow, Nauka, 1981; pp. 281-282.

[9]     Margrelashvili NZ, Vlasov P. Native growth inhibitor of a tea plant. Subtropicheskie kultury, 1986; 4:81-85.

[10]    Kefeli VI, Chailakhyan M. New tendencies in the study of plant growth regulators. Uspekhi Sovremennoi Biologii, 1975; 80: 116-127.

[11]    Wareing PF, Phillips LD. The Control of Growth and Differentiation in Plants. Oxford, Pergamon Press 1978; pp. 348.

[12]    Kefeli VI, Lozhnikova VN, Kblopcnkova LP. Activity of phytohormones and native inhibitors in pea mutants being different in stem height. Izv Akad Nauk SSSR. Serbia, 1973; 5: 681-687.

[13]    Chemobrovkina, NP Kefeli VI, Ivanova, RF. Dormancy, Growth and native regulators in buds and seeds of kareliya-birch. Fiziologiya rastenii. 1975; 22: 1013-1019.

[14]    Nemsadze NP, Bagrationi NN. Activity of endogenous growth substances in maize seedlings which differ in growth intensity. In: Trudy Tbilisskogo Univ., 1979; 199: 110-115.

[15]    Dzagnidze SH, Chasnishvili SH. Dynamic of activity of stimulators, inhibitors and gibberellin-like compunds in the shoots of grape cv. Rkatsiteli. In: Plant Growth and Development Regulators. Moscow 1981.

[16]    Terek OI. Endogenous auxins and gibberellins of bean plants under the effect of physiologically active compounds. In: Plant Growth and Development regulators, Moscow, Nauka. 1981

[17]    Kefeli VI, Kislin E Gas-chromatographic assay of abscisic and indolyl 3-acetic acids in plant tissues. Fiziologiya Rastenii, 1982; 23: 407-413.

[18]    Tkhi-Muoi L. Growth and regeneration ability of a plant and its isolated parts in connection with the effect of phytohormones and inhibitors. Dr. Sci. Thesis. Moscow. Institute of Plant Physiology. 1985.

[19]    Taniya LI, Yusufova MH, Tkhi-Muoi L, Kefeli VI. Inhibitory effect of humic acids and coumarin on the process of root formation in the bean leaf and stem cuttings. Fiziologiya rastenii. 1986; 33: 951-954.

[20]    Muzafarov EN, Zolotareva EK. Uncoupling Effect of Hydrocinnamic Acid Derivatives in Pea Chloroplasts. Biochem. Physiol. Pflanzen. 1989; 184: 363-369.

[21]    Nazarova GN, Lubinov VY, Muzafarov EN. Effect of quercetin, rutin on the activity of NADP-dependent glycerol-aldehide phosphate dehydrogenase complex. Biokhimiya, 1989; 54: 85-856.

[22]    Kvartshava LS, Kefeli VI. Effect of light of various intensities on the formation of amaranth-type in amaranth cotyledons. Fiziologiya rastenii. 1980; 27: 423-424.

[23]    Aknazarov OA, Shomansurov S. Changes in the content of endogenous growth regulators when the plants are transferred to various altitudes in the Pamirs. In: Plant Growth and Development Regulators. Moscow 1981; pp. 60-68.

[24]    Makarova RV, Paskual EV, Martynish F, Sanches P, Ranavira K. Effect of 6-benzylaminopurine on the growth of soybean cotyledons and hypocotyl. Biologicheskie nauki. 1988; 8: 81-86.

[25]    Rakitina TY, Kefeli V1. Ethylene evolution and pea plant growth. Fiziologiya rastenii, 1989; 36: 789-793.

[26]    Allakhverdiev SL, Muzafarov EN, Klimov V. Quercetin effect on electron transfer in photosystems I and II of pea chloroplasts. Biofizika, 1989; 34: 976-979.

[27]    Kefeli VI, Kof E. Growth inhibitors of phenolic nature: Some aspects of formation and action. In: Plant Growth Regulators, Sofia. 1983; pp. 283-288.

[28]    Kof EM,, Kefeli V1. Conjugated phenolic compounds in green and dark grown mutants of pea plants conjugated plant hormones structure, metabolism and function. DDR Halle, 1987; pp.361-368.

[29]    Kof EM, Sharipov GD, Kutacek M, Vassilev G, Kefeli VI Hormonal regulation of tall and dwarf pea plant growth. In: Plant Growth Regulators, Sofia, 1983; pp. 298-301.

[30]    Kefeli VI. Natural Plant Growth Inhibitors and Phytohormones. The Hague Boston: Dr. W. Junk Publishers 1978

[31]    Kefeli Vl, Kutacek M, Vaskova K. Influence of natural substances of phenolic character and diethyldithiocarbamide on the metabolism of tryptophan in cabbage, maize and pea. Biologia Plantarum 1970; 2: 90.

[32]    Amrhein N, Godeke J, Gerhardt, Kefeli V Die bestimmung der aktivitiit der phenylalanin ammonium-lyase in intahten pflanzenzellen. Tagung der Deutschen Botanischen Geselschaft, Zurich 1976.

[33]    Turetskaya R, Kefeli VI, Kutacek M, Tshimakovski M, Krupnikova T. Isolation and some physiological properties of natural plant growth inhibitors. Biologia Plantarum l968; 10: 205-209.

[34]    Kefeli VL, Ranaweera K, Piruzian ES, Makarova RV, Andrianov VM, Ragubova S, Yusibov VM. Phytohormones in transformed tobacco plants. In: Molecular Aspects of Hormonal Regulation. The Hague, the Netherlands: SPB Acad. Publ. 1990; pp. 137-144.

[35]    Polyakov AS, Kof EM, Gostimskii SA. Phytohormones and pigments in albino mutants. Biologia Plantarum 1985; 27: 139-144.

[36]    Ragubova SD. Effect of growth regulators on the yield of green and albino plants. In: Abstracts II Konf, Nauchn.-Uch. Tsentra UDN, Moscow, 1989; 201-202.

[37]    Nadzhimov UK, Abzalov MF. Cotton dwarfs and physiologically active substances. In: Growth Regulators and Plant Development, Moscow, Nauka 1988; pp.124.

[38]    Lev SV. Tissues and Organs of Cotton Plants as Biotests for Determining Endogenous Growth. Ph. D. Thesis. Moscow. 1983.

[39]    Kefeli VI, Pirusian ES, Makarova RV, Ranavira K, Andrianov VM Jusibo VM. Phytohormones and phenolics in transformed tobacco plants. - In: 6th Congress of the Federation of European Societies of Plant Physiology (FESPP)/Split-Yugoslavia. 4-10 September 1988.

[40]    Saidova SA. Characteristic inhibitory effect of some native and synthetic inhibitors on plant growth. Ph. D. Thesis. Moscow. (1972)

[41]    Umarov AL. Benzimidazole derivatives-regulators with double effect. In: Plant Growth and Development Regulators, Moscow, Nauka, 1981; pp. 87.

[42]    Arapetyan FR, Muromtsev V, Koreneva VM. Effect of Fusicoccin on the Growth and Length of Anthocyanic Zone of the *Zea Mays Root* and Seedling. In: Plant Growth and Development Regulators. Moscow, 1981; 101-113.

[43]    Shapkin VA. Some peculiarities of physiologic effect of benzimidazole derivatives Moscow, Nauka, 1981; 2: 2-213.

[44]    Soma S, Kefeli VI. Effect of phytohormone concentration on callus formation in tomato plants. In: Abstracts II Konf. Nauchn.Uch. Tsentra UDN, Moscow, 1989; pp.218-219.

[45]    Soma S, Ezhoa TA, Kefeli VI, Gostimsky SA. Cytokinin and pigment analysis of regenerated tomato plants *in vitro*. In: Abstr. 7 Intern. Congress of Plant Tissue and Culture, Amsterdam, 1990; pp.280-281.

# CHAPTER 13

# Secondary Substances for Chemo-Taxonomy and Allelopathy

**Abstract:** Secondary substances are the products of photosynthesis which form in plants from primary products like sugars and amino acids. Among them are alkaloids, terpenoids, phenolics, lignins and alkaloids. Plant hormones are also members of secondary substances. Some of them play an important role in the allelopathic interactions between the plants in the eco-cenosis.

## INTRODUCTION

The difficulty in obtaining comprehensive chemical information of the type necessary for phylogenetic comparisons between large numbers of plant species proved to be almost impossible until the development of modern physical methods in organic chemistry. Yet despite the great difficulty in chemically characterizing taxa in the early chemotaxonomic surveys, these early contributors must be credited with providing a philosophical foundation for the application of modern chemistry to taxonomy [1].

Chemical surveys of taxonomically more restricted groupings of plant species have been more successful in providing data for the working phylogenist. The distribution of betalains, formerly known as nitrogen-containing anthocyanins, along with other evidence suggest that the Cactaceae should be aligned with the Centrospermae and this disposition is now generally accepted from an extensive survey of anthocyanin content in plant species. It was observed also that the incidence of pelargonidin pigmentation is more common in tropical species than in those from temperate regions. They also noted that the flowers of the majority of tropical or subtropical species containing cyanidin or delphinidin were appreciably redder than temperate species with the same anthocyanins in their flowers. They concluded that natural selection favors the survival of red-flowered forms in the tropics.

It is now well known that pollination vectors which favor red are common in tropical regions. Anthocyanins are able to form in the light. Thus growth of roots and shoots of wheat and corn was investigated under dark and light conditions. It was observed that corn roots growth is depressed by the white light. The light source is luminescent lamps. No anthyocyanins were formed during the darkness in the corn roots. The length of the red stained of roots corresponds to differentiation region of the roots. In wheat seedlings anthocyanin was not accumulated during the darkness. After the light illumination, only coleoptiles became red. The appearance of anthocyanins in the coleoptile of field corn Carolina is observed after one hour of illumination of seedlings. Two dimensional chromatography allowed observing only one anthocyanin in wheat coleoptile and in the corn roots. Mobility of anthocyanin on the Watmann No. I paper was Rf 0.46 (solvent: butanol: acetic acid: water, 4:1:2). The same stain in the 15% acetate was more mobile and had Rf - 0.9. UV spectra of anthyocyanins show their spectra data with two max at 3,506 nm and 311 nm, which corresponds to anthocyanin picks.

The extensive survey of a very large number of dicotyledonous and monocotyledonous species for flavonoid and cinnamic acid content by Zaprometov [2] represents a good example of a survey for the distribution of a widely occurring compound grouping for phylogenetic comparison purposes. The classes of flavonoids include the catechins. XV; leucoanthocyanidins. XVI; flavones, IX; flavonols, X; anthocyanidins, VI-VIII; flavanones. XVII; flavanols, XVIII; chalcones, XIX; dihydrochalcones, XX; aurones, XXI; and isoflavones. XXII. Two common cinnamic acids are p-coumaric acid, XXIII, and caffeic acid, XXIV. Some phytochemical correlations may be noted from Bate-Smith's survey of the angiosperm polyphenols. In the dicots, nearly all of the predominantly woody families from Casuarinaceae to Ebenaceae contain leucoanthocyanidins, while nearly all of the predominately herbaceous families are from the Aristolochiaceae.

Phenolic natural inhibitors suppress extension of coleoptile segments, growth of epicotyls and bud breaking. While this suppression is only temporary, it is sufficiently strong. Whereas processes that occur within the meristem are independent of natural inhibitors, cell extension processes are sensitive to these substances.

It can be assumed that in autumn buds grow vigorously and embryos form within the seeds. But their further growth and extension are predominantly blocked by inhibitors. In spring, when the levels of inhibitors decrease and the

Narcin Palavan-Unsal (Ed)

concentration of stimulating hormones increases, embryonic organs begin growing intensively. Artificial application of natural growth inhibitors into either cuttings with opening buds or into germinating seeds inhibits extension processes. It is possible to observe the rise of growth-inhibiting substances within vegetative pea plants when the process of growth is depressed by either mutagens or light [3]. A dramatic increase in p-coumaric acid in both free and bound quercetin (glucosil coumarate (QGC) occurs. Bioassay of p-coumarate and QGC, utilizing either wheat or corn coleoptile or, pea stem sections, reveals that 175 mg/l p-coumaric acid inhibits the elongation of wheat coleoptile sections, but that the elongations of *Pisum* stem and *Zea* coleoptile sections are inhibited at 700 mg/1. Comparison of the lengths and weights of stem sections of *Pisum* and *Zea* coleoptile sections incubated with p--coumaric acid shows that both are similar in their sensitivities to this inhibitor. Analysis of the sensitivity of wheat coleoptile and *Pisum* stem sections to QGC leads to the conclusions that the inhibition of elongation of the former begins at 4000 mg/l, while the elongation of the latter is inhibited only at 8000 mg/l. Thus, in this case, plant-donor sections are also less sensitive than wheat coleoptile sections. Phenolics can also secrete from roots and play role of the allochemical agents.

Considering the taxonomic importance of instances of compounds with unique distributions in plants, the study of the "nitrogenous anthocyanins" is perhaps the best known example. These red pigments were first recognized as different from the anthocyanins by Schudel in 1918 but the structures were not fully resolved until 1962. The betalains, as they are now known, include the red to violet betacyanins and the yellow betaxanthins. These compounds are distributed only in certain plant families, and this has been recognized for some years as being of importance in establishing the natural relationships of families in the Centrospermae.

## COMPOUND GROUPINGS COMMENTS ON BOTANICAL DISTRIBUTION

Some major groupings of plant natural products are listed along with some comment on their botanical distribution. This brief summary is only intended to give some insights into the scope of "natural products chemotaxonomy" (Table **1**).

Thus, plants contain hormones, secondary substances of purine, indolic and fluorine nature. Besides that, plants contain growth regulators that are non-hormonal in nature. These regulators change in concentration during ontogeny and when applied exogenously, can either stimulate or depress growth. The bulk of the phenolic or terpenoid regulators are localized. There is lack of data regarding compartmentalization of many of the inhibitors. This raises the question of whether their intracellular concentrations become elevated sufficiently to affect metabolic pathways *in vivo*. Exogenously applied regulators of non-hormonal nature usually interfere with growth only at high concentrations. Therefore, this possibility cannot be excluded that under these conditions, reactions occur within the cell that are absent *in vivo*.

As previously mentioned, this large class of phenolic compounds possesses substances that are either protectors or synergists. The phenolics include growth inhibiting substances. Although they do not resemble each other structurally, they possess similar biological properties. In addition to the phenolics of higher plants which possess inhibitory properties, lower organisms also contain some inhibiting compounds. Lunularic acid, a dihydrostilbene carboxylic acid, has been found in all liverworts and algae. Thus far examined for its presence these lower plants do not contain ABA, the ubiquitous function in lower plants that ABA does in higher plants. Lunularic acid is therefore a plant growth regulator of some chemotaxonomic and phylogenetic significance. Natural phenolic inhibitors do not depress all growth processes, mainly they block cell elongation.

Elongation of coleoptile segments and root formation in *Phaseolus* cuttings differ in their responses to the metabolic inhibitors, although both tissues display a high sensitivity to indole-3 acetic acid (IAA). Coleoptile segments that have been treated with IAA are insensitive to low threshold doses of nucleic-and protein-metabolism inhibitors, and also to inhibitors of photosynthetic phosphorylation, while they are highly sensitive to dinitrophenol. In contrast, *Phaseolus* cuttings have low sensitivity to dinitrophenol, but their rhizogenesis is strongly suppressed by the metabolic inhibitors. Cell division differs from cell extension in its reaction to natural inhibitors. Unlike cell extension, cell division processes are not suppressed and sometimes they are even slightly activated in willow in the presence of natural phenolic inhibitors. Cell extension is extremely sensitive to phenolic inhibitors, particularly to ABA.

**Table 1:** A summary of some groupings of natural products which have been used in chemotaxonomic correlation including some comments on the botanical distribution of these constituents.

| Compound Grouping | Comments on Botanical Distribution |
|---|---|
| Terpenoids | Widely distributed |
| Monoterpenes | Widely distributed of taxonomic value primarily below the generic level |
| Sesquiterpenoids | Rather wide distribution, but particularly important and useful in the taxonomy of the Compositae |
| Diterpenes | Rather wide distribution in the seed plants, taxonomically useful in specific groups such as conifers |
| Tritepenes | Wide distribution in living organisms, taxonomically useful in several angiosperm families such as Cucurbitaceae and in the fungi. |
| Carotenoids | The universally distributed photosynthetic caretonoids useful in algal classification |
| Flavonoids | Very large number of diverse compounds found throughout the vascular plants and in nearly, if not all, angiosperms of great taxonomic value below the genus level and of possible use in the classification of higher categories. |
| Lignins and lignans | Useful in the classification of higher categories. |
| Quinones | Widely distributed among living organisms, but compounds of this type with limited distribution have potential value in the classification of a number of angiosperm families. |
| Polysaccharides | Universally distributed, probably useful in the classification of higher categories, particularly among the algae, comparative data mostly lacking. |
| Plant glycosides | Great chemical variations in the non-sugar portion of the molecule; varyingly of the great usefulness in taxonomy |
| Asperulosides and Acubins (Iridoid glycosides) | Rubiaceae, Scrophulariaceae and related families, Plantaginaceae, Comaceae and others. |
| Ranunculins | Found only in Ranunculaceae |
| Cyanogenetic compounds | Ranunculaceae and other plant families |
| Polyalcohols | Widely distributed but have taxonomic potential |
| Sulphur compounds | A chemically diverse grouping of compounds which in their various forms are widely distributed the isothiocyanate-producing glycosides are characteristic of families in the Rhoedales. |
| Amino acids (non-protein) | Liliaceae and related families. Leguminosae (particularly Papilionatae and other groups) |
| Alkaloids | A chemically and biosynthetically diverse group of compounds rarely found in the lower vascular plants and irregularly distributed among the angiosperms; highly useful in the taxonomy of some groups. |
| Betacyanins and betaxanhins | Sentrospermae 'Betanales'. |
| Alkanes (fatty acids and waxes) | Wldely distributed, possible of us, in classification below the genus level and of use in organic geochemistry |
| Fatty acid epoxides | Found in seven families |
| Acetylenes | Distributed in the basidiomycetes and at least 13 angiospermae families of particular use in the link between the Compositae and Umbellifera. |
| Assorted compounds | Ferns |

There are several newly discovered natural inhibitors of non-phenolic nature in recent years. A group of triterpenoids, cucurbitacins, have been found within cucumbers, melons, squashes, and pumpkins. These substances appear to function as anti-gibberellins and also they inhibit the elongation of the second internode on bean bioassay. Cucurbitacins are composed of similar substances.

Other inhibitors include diterpenes containing aperhydrozulene nucleus, e.g. portual is a typical antiauxin. Investigations of batatasins have yielded considerable information on the mode of action of non-hormonal terpenoid inhibitors. Batatasins were isolated from dormant Chinese yam bulbils (*Dioscorea batatas*). They inhibit the dormancy of bulbils and influence auxin transport. The synthetic analogues of batatasins have now been prepared. Podolactone depresses the mitotic process as well as elongation. Extraction and identification of natural growth inhibitors have progressed more rapidly than physiological and biochemical characterization of their actions. More recently terpenoid inhibitors were discovered include those that affect cell elongation (lycoricidin from *Lycoris* and *Narcissus*, and heliangine from *Helianthus*) seed germination and stem growth inhibitors (cucurbic acid from *Cucurbita pepo*). In addition to these compounds, the ABA metabolites, ABA-glucose esther, metabolite C, phaseic acid and 4-dihydrophaseic acid may inhibit various plant growth and developmental processes. It can also play a role of the allelopathogens. They can also be found within other cellular compartments where they may act upon metabolic pathways, modifying either cell multiplication or elongation.

Non-hormonal growth regulators may affect the synthesis and/or destruction of phytohormones, mainly IAA. These regulators behave non-specifically, modifying the actions of auxins, gibberellins and cytokinins upon growth. A variety of both uncertainties and unresolved contradictions exist that have prevented a thorough elucidation of the mechanisms of action of both phenolic and terpoid regulators.

## REFERENCES:

[1]    Harborne JB. Plant Phenolics. Encycl Plant Physiol Vol. 8. Bell E and Charlwood BV Eds. Berlin, 1980; pp. 392-402.
[2]    Zaprometov MN. Osnovy biohimii fenol'nyh soedinenii (The Basic Biochemistry of Phenolic Compounds), Moscow, Vysshaya Shkola, 1974.
[3]    Kefeli VI. Natural Plant Growth Inhibitors and Phytohormones. W Junk Ed. BV Publishers. Boston, 1978; pp. 277.

# CHAPTER 14

## Some Characteristics of Willow Species for their Taxonomy

**Abstract:** Some plant species have not only morphological but also physiological peculiarities. When it is difficult to determine the species by morphologic characteristics, it is possible to use physiological and biochemical parameters. Thus, different willow species have different reactions of auxin during rooting. Other species have different reactions during dormancy process. These reactions are very important for the selection of willow species for the propagation on the fabricated soil.

## INTRODUCTION

These physiological characteristics rooting bud opening, and reaction to auxin, could be used as taxonomic features for willow identification. The rooting of poplar was connected with the position of cutting on the mother plant. The highest intensity of rooting was specific for the basal part of the twig of poplar *(Populus nigra)*.

Cuttings of six willow species were rooted in water (control) and after auxin (indole-3-acetic acid, 150 mg/l) treatment. The intensity of the dormancy was determined by the following data: root formation, shoot development, and reaction on auxin. Silky, weeping, and pussy willows do not have deep dormancy and were sensitive to auxin. The cuttings of purple, autumn and white willows were less active in the rooting process and less sensitive to auxin. These species do not open their buds in favorite laboratory conditions.

There are many external factors, which have an influence on the regulation processes in the developing physiological processes [1-3]. In our experiments, the main process of willow and poplar propagation by cuttings depends on the time of the rooting of the cuttings and the position of the cutting on the twig of the mother plant. There are many internal factors, which had an influence on the rooting processes, including active wood formation, dormancy factors, phenolic accumulation and absence of the rooting hormones, auxins [4, 5]. It was very important to determine the effects of auxin on the rooting of different willow species for vegetative propagation.

Mother plants of willows and poplar were donors of cuttings and grown at the Jennings Environmental Education Center (JEEC) in Slippery Rock, PA. The mother plants were grown on FS [6]. The cuttings, 18 cm long, were treated by indole acetic acid (IAA), 150 mg/l. The rooting process proceeded in the lab at 25°C during 2 weeks. We selected the time (October 2004) when growth of the mother shoots stops and dormancy develops.

Plant growth is a process, based on the rhythmic phenomena. In our experiments with willow and poplar shoots in JEEC, we observed the dependency of the shoot elongation process with wood formation. The process of stem elongation of *Populus nigra* and *Salix rubra* was more intensive in the period of May through June. In July, the process of stem elongation was inhibited and the process of differentiation of wood activated.

Six species of willows were compared on their rhythmic of shoot growth. They differ in the process of growth rhythmic inhibition (GRI). White willow was a species with a very intensive GRI while purple willow had a low GRI. The other species, silky, pussy, weeping, and autumn willows, occupy an intermediate position.

GRI is a process which is accompanied by the dormancy process and develops with different intensity for various willow species.

Willows were used as donors of cuttings. The experiment was carried out in the fall when rooting is usually not very active (Table **1**).

Two willow species, *Salix sericea* and *Salix discolor,* could be used for rooting in the fall because they do not have dormancy. The other willows are dormant and do not form roots or develop shoots.

**Table 1:** Rooting of the cuttings of 6 willow species *(Salix spp.)* Experiment started on September 25, 2004, and took place at Alter Lab. Jennings Environmental Education Center. Data was recorded on October 1, 2004. C. Control; A: Auxin, indole-3-acetic acid, 150 mg/l; R: Amounts of root per cuttings; S. Amount of shoot per cuttings; R/S: Ratio of amounts of roots to shoots

| Name | Sample | R | S | R/S |
|---|---|---|---|---|
| | 1 C | 1.40 | 0.75 | 100/100 |
| | 1 A | 4.00 | 0.25 | 285/36 |
| *S. sericea* | 2 C | 1.50 | 1.80 | 100/100 |
| (Marsh silky) | 2 A | 3.70 | 0.30 | 246/17 |
| | 3 C | 4.30 | 1.30 | 100/100 |
| | 3 A | 4.10 | 0.40 | 95/31 |
| | 1 C | 0.64 | 0.57 | 100/100 |
| | 1 A | 1.21 | 0.36 | 189/61 |
| *S. babylonica* | 2 C | 2.60 | 1.00 | 100/100 |
| (L weeping) | 2 A | 3.50 | 0.07 | 134/7 |
| | 3 C | 5.07 | 1.20 | 100/100 |
| | 3 A | 7.90 | 0.42 | 158/35 |
| | 1 C | 0.72 | 0.16 | 100/100 |
| | 1 A | 0.42 | 0.57 | 58/100 |
| *S. rubra* | 2 C | 0.59 | 0.31 | 100/100 |
| (L purple) | 2 A | 0.40 | 0.25 | 67/81 |
| | 3 C | 1.45 | 0.13 | 100/100 |
| | 3 A | 0.75 | 0.35 | 51/269 |
| | 1 C | 1.00 | 0.66 | 100/100 |
| | 1 A | 4.00 | 0.17 | 400/25 |
| *S. discolor* | 2 C | 1.40 | 1.30 | 100/100 |
| (Muhl pussy) | 2 A | 2.16 | 0.0 | 154/0 |
| | 3 C | 2.90 | 1.25 | 100/100 |
| | 3 A | 0.08 | 0.16 | 2.7/12 |
| | 1 C | 0.38 | 0.30 | 100/100 |
| | 1 A | 0.06 | 0.0 | 15/0 |
| *S. senissima* | 2 C | 0.21 | 0.35 | 100/100 |
| (Fern autumn) | 2 A | 0.50 | 0.0 | 238/0 |
| | 3 C | 0.41 | 0.53 | 100/100 |
| | 3 A | 0.81 | 0.10 | 191/18 |
| | 1 C | 1.00 | 0.0 | 100/100 |
| | 1 A | 0.55 | 0.0 | 55/0 |
| *S. alba* | 2 C | 0.37 | 0.0 | 100/100 |
| (L white) | 2 A | 1.66 | 0.0 | 448/0 |
| | 3 C | 0.73 | 0.0 | 100/100 |
| | 3 A | 1.73 | 0.0 | 236/0 |

Auxin IAA, stimulates the rooting of the cuttings of silky and pussy willows but does not have an effect on the rooting of the cuttings of white, autumn, weeping and purple willows.

The second experiment confirms that auxin has a stimulating effect on the fall rooting of the cuttings of silky willow and pussy willow (Table **2**).

**Table 2:** Rooting of the cuttings of 6 willow species *(Salix spp).* The second experiment started on October 12. 2004 and took place at Alter Lab, Jennings Environmental Education Center. Data was recorded on October 26, 2004. C. Control; A: Auxin, indole-3-acetic acid, 150 mg/l; R: Amounts of root per cuttings; S. Amount of shoot per cuttings; R/S: Ratio of amounts of roots to shoots.

| Name | Sample | R | S | R/S |
|---|---|---|---|---|
| *S. sericea* | 1C | 3.60 | 1.40 | 100/100 |
| (Marsh, silky) | 1A | 11.50 | 0.87 | 319/62 |

**Table 2: cont....**

| | | | | |
|---|---|---|---|---|
| *S. babylonica* | 1C | 3.71 | 0.20 | 100/100 |
| (L. weeping) | 1A | 3.63 | 0.30 | 97/150 |
| *S. rubra* | 1C | 0.83 | 0.0 | 100/100 |
| (L. purple) | 1A | 4.63 | 0.0 | 0.0 |
| *S. discolor* | 1C | 0.70 | 0.22 | 100/100 |
| (Muhl, pussy) | 1A | 2.37 | 0.75 | 338/340 |
| *S. senissima* | 1C | 0.0 | 0.0 | 0.0 |
| (Fern, autumn) | 1A | 0.0 | 0.0 | 0.0 |
| *S. alba* | 1C | 0.0 | 0.0 | 0.0 |
| (L. white) | 1A | 0.0 | 0.0 | 0.0 |

Different willow species have a specific reaction on IAA in the case of bud opening. Spring cuttings of weeping willow had less open buds than the fall ones. The other species have more open buds in the spring. IAA stimulates bud opening for pussy willow in the fall whereas the other species have the opposite reaction (Table **3**).

**Table 3:** Comparative data on the rooting of willow cuttings from two experiments started on September 25, 2004 and October 12. 2004. C. Control; a: Auxin, indole-3-acetic acid, 150 mg/l; R: Amounts of root per cuttings; S. Amount of shoot per cuttings; R/S: Ratio of amounts of roots to shoots.

| Name | Sample | R | S | R/S |
|---|---|---|---|---|
| | Exp. 2 C | 3.60 | 1.40 | 100/100 |
| *S. sericea* | 2 A | 11.50 | 0.87 | 319/62 |
| (Marsh, silky) | Exp. 1C | 1.40 | 0.75 | 100/100 |
| | 1 A | 4.00 | 0.25 | 285/36 |
| | Exp. 2 C | 3.71 | 0.20 | 100/100 |
| *S. babylonica* | 2 A | 3.63 | 0.30 | 97/150 |
| (L. weeping) | Exp. 1C | 0.64 | 0.57 | 100/100 |
| | 1 A | 1.21 | 0.36 | 189/61 |
| *S. rubra* | Exp. 1C | 0.72 | 0.16 | 100/100 |
| (Red willow) | 1 A | 0.42 | 0.57 | 58/100 |
| | Exp. 2 C | 0.70 | 0.22 | 100/100 |
| *S. discolor* | 2 A | 2.37 | 0.75 | 338/340 |
| (Muhl, pussy) | Exp. 1C | 1.00 | 0.66 | 100/100 |
| | 1 A | 4.00 | 0.17 | 400/25 |
| | Exp. 2 C | 0.0 | 0.0 | 0.0 |
| *S. senissima* | 2 A | 0.0 | 0.0 | 0.0 |
| (Fern, autumn) | Exp. 1C | 0.38 | 0.30 | 100/100 |
| | 1 A | 0.06 | 0.0 | 15/0 |
| | Exp. 2 C | 0.0 | 0.0 | 0.0 |
| *S. alba* | 2 A | 0.0 | 0.0 | 0.0 |
| (L. white) | Exp. 1C | 1.00 | 0.0 | 100/100 |
| | 1 A | 0.55 | 0.0 | 55/0 |

Willow species have different capabilities of rooting and shooting in our experiments when plants grow on fabricated soils [7, 8].

Some species like autumn willow are not able to form roots and shoots on their cuttings in the fall or in the spring. Some, like white and red willows, do not open buds on their cuttings in the fall (Table **4**). Pussy willow is able to form roots and open buds in the fall and in the spring. Pussy and weeping willows have a positive reaction (active rooting) to the auxin (IAA, 150 mg/l). It confers to the theory of hormone - inhibitors relations [9, 10]. Thus, such differences in the growth and regeneration properties could be used as additional characteristics for willow taxonomy.

**Table 4:** Rooting and bud opening of willow cuttings in the fall 2004 (F) and in the spring 2005 (S).

| Samples of willow | F or S | Roots on one cutting | | | Open buds on one cuttings | | |
|---|---|---|---|---|---|---|---|
| | | Water | IAA | IAA/water ratio (%) | Water | IAA | IAA/water ratio (%) |
| Silky | F | 10.1 | 15.3 | 151 | 1.8 | 1.3 | 72 |
| | S | 7.1 | 5.8 | 81 | 4.0 | 2.8 | 70 |
| Autumn | F | 0.0 | 0.0 | 0.0 | 0.0 | 0.0 | 0.0 |
| | S | 0.0 | 0.0 | 0.0 | 0.0 | 0.0 | 0.0 |
| Red | F | 1.8 | 15.3 | 151 | 0.0 | 0.2 | 0.0 |
| | S | 6.6 | 5.7 | 86 | 2.4 | 1.9 | 79 |
| White | F | 0.8 | 1.8 | 225 | 0.0 | 0.0 | 0.0 |
| | S | 5.2 | 9.0 | 173 | 1.6 | 1.9 | 118 |
| Weeping | F | 8.9 | 9.6 | 107 | 8.9 | 9.6 | 107 |
| | S | 5.2 | 9.0 | 173 | 0.8 | 1.1 | 132 |
| Pussy | F | 2.2 | 8.5 | 386 | 0.3 | 1.1 | 300 |
| | S | 4.9 | 11.7 | 238 | 1.7 | 1.8 | 107 |

# REFERENCES

[1]    Ozalpan A. Basic Radiobiology, published in Golden Horn University, Istanbul. 2001; pp. 353

[2]    Palavan-Unsal N, Cag S, Cetin E, Buyuktuncer D. Retardation of Senescence by meta-Topolin in Wheat Leaves. Journal of Cell and Molecular Biology. Golden Horn University, Turkey. 2002; 1: 101-108.

[3]    Palavan-Unsal N. Book Review. Natural Growth Inhibitors in Plant and Environment. Journal of Cell and Molecular Biology. Golden Horn University, Istanbul, Turkey. 2003; 2: 60.

[4]    Turetskaya R. Physiology of Rootings in Cuttings (in Russian). Moscow, Nauka. 1961.

[5]    Turetskaya R, Kefeli V, Kof E. Role of Natural Regulators in Ontogenesis. J Sov Plant Physiol. 1966; 13: 29.

[6]    Kalevitch MV, Kefeli VI. Fungi in fabricated soil. Int J Environ and Pollution 2007; 29: 424-434.

[7]    Kefeli VI. Water cleaning and wetland construction. Int J Environ and Pollution 2007; 29: 383-391.

[8]    Kefeli V, Dunn MH, Johnson D, Taylor W. Fabricated soils for landscape restoration: An example for scientific contribution by public-private partnership effort. Int J Environ and Pollution 2007; 29: 405-411.

[9]    Kefeli Y, Kadyrov C. Natural Growth Inhibitors. Their Chemical and Physiological Properties. Annual Review of Plant Physiolology. 1971; 22: 185-196.

[10]   Tretjakov NN. Plant Physiology (Russ.) Moscow Agropromizdat 2002.

## CHAPTER 15

# Relations of Growth and Differentiation of Willow and Poplar Species: Biological Rhythms on Fabricated Soils

**Abstract:** The main factor of plant growth is periodicity- rhythms of stem elongation. We selected fast growing willow and poplar forms for the propagation on the fabricated soils. Therefore, we determined the types of stem growth of willows and compared these processes with the stem differentiation processes. This last process is connected with the lignifications of stem cells.

## INTRODUCTION

The growth process of shoots proceeds in a rhythmic manner. Each element of shoot elongation is accompanied by leaf morphogenesis and bud formation. Sabinin [1]. and his students were the first researchers to investigate the phenomena of shoot elongation. Each wave of growth is accompanied by a peak and followed by a gap in the growth process. The researchers [2] investigated a more simple type of growth rhythmic in wheat coleoptile sections. Here we present a more complex process dealing with the differentiation of the stem conducting tissues and leaf-bud development process (plastochron) on the one-year shoots which stopped their growth. The mother plants were grown on FS [3].

Willow and poplar plants were grown at Jennings Environmental Education Center (JEEC) on the fabricated soil (FS) plot. The one-year shoots were cut from the mother plants, while the internodes from the base to the top of the shoot were measured in cm. The data was plotted into graphs and the rhythmic growth of the shoot determined (Table **1**).

**Table 1:** Rhythmic growth of poplar and willow shoots (internode length in cm). Shoot growth rhythm for poplar and willow shoots during the period of intensive growth, May and June 2004.

| Plant | Internode Length (from basal part to apex, cm) |
|---|---|
| Poplar | 3.59 (average of 63 measurements) |
| Red willow | 5.15 (average of 47 measurements) |
| Pussy willow | 3.96 (average of 47 measurements) |

Poplar and willow shoots have different lengths of internodes during their growth. They possess rhythm in the process of elongation. Shoots of six willow species were cut from the main stem and measured in September and October 2004, while their growth was stopped. Distance from base (DFB) and internode length (IL) were measured for four shoots (Table **2**).

**Table 2:** Data for the white willow (*Salix alba* L.) shoots.

| | Number 1 | | Number 2 | | Number 3 | | Number 4 |
|---|---|---|---|---|---|---|---|
| DFB | IL | DFB | IL | DFB | IL | DFB | IL |
| 2.2 | 0.1 | 1.7 | 0.4 | 1.9 | 1.1 | 1.2 | 0.9 |
| 2.3 | 1.5 | 2.1 | 1.2 | 3 | 1.6 | 2.1 | 1 |
| 3.8 | 0.2 | 3.3 | 2 | 4.6 | 1.4 | 3.1 | 1.4 |
| 4 | 1.5 | 5.3 | 1.3 | 6 | 2.8 | 4.5 | 2 |
| 5.5 | 1.8 | 6.6 | 2.4 | 8.8 | 2.8 | 6.5 | 2.2 |
| 7.3 | 1.2 | 9 | 2.1 | 11.6 | 2.6 | 8.7 | 1.9 |
| 8.5 | 1.9 | 11.1 | 1.4 | 14.2 | 0.9 | 10.6 | 1.8 |
| 10.4 | 2.2 | 12.5 | 2 | 15.1 | 3.7 | 12.4 | 2.5 |
| 12.6 | 2 | 14.5 | 1.9 | 18.8 | 2.1 | 14.9 | 2.7 |
| 14.6 | 1.9 | 16.4 | 0.7 | 20.9 | 0.2 | 17.6 | 2.4 |
| 16.5 | 2.7 | 17.1 | 2 | 21.1 | 2.8 | 20 | 2.4 |

**Table 2: cont....**

| | | | | | | | |
|---|---|---|---|---|---|---|---|
| 19.2 | 1.4 | 19.1 | 0.6 | 23.9 | 1.9 | 22.4 | 1.9* |
| 20.6 | 1.1 | 19.7 | 1.5 | 25.8 | 1.9 | 24.3 | 1.6 |
| 21.7 | 0.4 | 21.2 | 2.5 | 27.7 | 1.9* | 25.9 | 2 |
| 22.1 | 2.9* | 23.7 | 1.2 | 29.6 | 1.4 | 27.9 | 1.3 |
| 25 | 1.6 | 24.9 | 0.7 | 31 | 1.8 | 29.2 | 1.8* |
| 26.6 | 1.7* | 25.9 | 1.4* | 32.8 | 1.3 | 31 | 1.4 |
| 28.3 | 1.5 | 27 | 1.1 | 34.1 | 1.1 | 32.4 | 0.9 |
| 29.8 | 1.5 | 28.1 | 0.9 | 35.2 | 0.9 | 33.3 | 0.7 |
| 31.3 | 1 | 29 | 1.2* | 36.1 | 0.6 | 34 | 0.7 |
| 32.3 | 1.5 | 30.2 | 1 | 36.7 | 0.6 | 34.7 | 0.7 |
| 33.8 | 1.1 | 31.2 | 0.8 | 37.3 | 0.8 | 35.4 | 1 |
| 34.9 | 1.1 | 32 | 0.7 | 38.1 | 0.7 | 36.4 | 0.9 |
| 36 | 1.2 | 32.7 | 1.1 | 38.8 | 0.6 | 37.3 | 0.8 |
| 37.2 | 1.1 | 33.8 | 0.8 | 39.4 | 0.5 | 38.1 | 0.8 |
| 38.3 | 1.2 | 34.6 | 0.8 | 39.9 | 0.5 | 38.9 | 0.9 |
| 39.5 | 1.1 | 35.4 | 1 | 40.4 | 0.4 | 39.8 | 0.9 |
| 40.6 | 1 | 36.4 | 0.9 | 40.8 | 0.3 | 40.7 | 0.8 |
| 41.6 | 1.3 | 37.3 | 0.8 | 41.1 | 0.5 | 41.5 | 1.1 |
| 42.9 | 0.9 | 38.1 | 0.7 | 41.6 | 0.5 | 42.6 | 1.2 |
| 43.8 | 0.7 | 38.8 | 0.7 | 42.1 | 0.6 | 43.8 | 1 |
| 44.5 | 0.7 | 39.5 | 0.7 | 42.7 | 0.5 | 44.8 | 1.1 |
| 45.2 | 0.5 | 40.2 | 0.9 | 43.2 | 0.6 | 45.9 | 0.7 |

The rhythm of white willow shoots growth is evidenced. The most intensive was the growth of the internodes at the base of the shoot and in the middle of the shoot.

Growth of the thin poplar shoots proceeds in the rhythmic form. Poplar shoots with a difference in the thickness have the most intensive growth in the middle of the shoot. Some of them continue an intensive growth even at the top (6 mm thick).

Poplar shoots with medium thickness also have a rhythmic style of growth, but the wave of the intensive growth was switched to the top of the shoot (22-34 internodes). The thick shoots also have the rhythmic type of growth, but the growth intensity was switched to the apex of the shoot (35-45 internodes). There is no correlation between bud formation and stem thickness (two types of differentiation) in the case of thin shoots. A similar absence of correlation between stem thickness and bud formation was observed for medium poplar shoots. There was no correlation between stem thickness and bud formation in the large poplar shoots. The length of the thin shoots does not correlate with increasing stem diameter. Similar observation: the length of the shoots does not correlate with stem thickness. The thick stems of poplar do not have a positive correlation with the length of the stem. The growth of the poplar stem proceeds in a pulsating style and the rhythms of growth were observed mostly at the 6th and 7th internodes of the shoots. In the case of the medium shoots, activation of the growth was observed for 4-5 internodes of the shoots. The largest shoots did not have a visible activation of growth in the frames of the 1-7 internodes.

Willows and poplars, as sanitary plants, are easily propagated and could be used for water cleaning [4]. Shoots of willow and poplar grow in the form of pulsations (rhythms). Bud formation and stem thickness accompanied the shoot growth. The rhythms of stem growth (elongation) mostly depend on the intensity of the stem thickness but not on the bud formation. Thus elongation of the stem, bud formation, and stem thickness were the three forms of shoot growth which were investigated. The correlations between these three types of processes were analyzed. The most visible was the correlation between stem growth and stem thickness. In general, photosynthesis of leaves had an influence on stem elongation and stem thickness more directly. However, the process of bud formation is more independent.

# REFERENCES

[1]     Sabinin DA. Physiology of Plant Development (in Russian). M Ussa Academy Publication, 1963.

[2] Kefeli VI, Kalevitch MV. Natural growth inhibitors and phytohormones in plant and environment. Dordrecht: Kluwer Academic Publishers. 2003; pp 49-64.

[3] Kefeli V, Dunn MH, Johnson D, Taylor W. Fabricated soils for landscape restoration: An example for scientific contribution by public-private partnership effort. Int J Environ and Pollution 2007; 29: 405-411.

[4] Kefeli VI. Water cleaning and wetland construction. Int J Environ and Pollution 2007; 29: 383-391.

# CHAPTER 16

## Phenolic Cycle in Plants and Environment

**Abstract:** This chapter combines the processes in plants photosynthesis, growth and differentiation with the processes of humus formation in the soil. Chloroplasts in leaves are responsible for phenolic products formation whereas soil micelles are the units, which could be considered as alumni-silicate matrix covered by humus envelope. The plant residues in the soil could participate in humus formation.

### INTRODUCTION

Phenolics are very stable products in plant organisms. Generally, they are characterized by a benzene ring and one hydroxyl group (-OH). They can be converted into lignin which is the main phenolic polymer in plants. Microorganisms break down these molecules and their fragments contribute to the mineralization of soil nitrogen and humus formation. Thus humus participates actively in fulfilling plants nutritional needs and growth. Light enhances the biosynthesis of phenolic substances in plant chloroplasts and these constitute in addition to soil micelles (humus) a second formation site for this diverse group of organic molecules. It should be mentioned however, that phenolics tend to accumulate in plant vacuoles in relatively high amounts, or they deposit in the secondary cell wall as lignin.

### CHLOROPLASTS AS CENTERS OF PHENOLICS BIOSYNTHESIS

Experiments with chloroplasts of willow (*Salix* spp.) leaves showed that the synthesis of phenolcarboxylic acids and flavonoids is strongly stimulated by light exposure. Metabolic inhibitors that depress photosynthetic activity (simazine, diurone, and chloramphenicol) affect negatively the biosynthesis of flavonoids. Leaves chloroplasts have the capability to localize phenol compounds, some of which are specific to these organelles only. The chloroplast of spring willow leaves contain more phenols than the chloroplasts of the same leaves in the autumn. Light is a mandatory condition to initiate phenolics synthesis and this is indicated also by the lack of such molecules in the protoplastids of etiolated willow shoots [1]. Light appears also to induce flavonols synthesis in the chloroplasts and cytoplasm. Chalcone and phenolcarbonic acid present in etiolated willow shoots can be considered metabolic precursors of light-synthesized flavonols. In certain cell compartments (vacuoles and cell wall) phenols are contained in significant amounts [2]. However, it is not clear yet how phenols are translocated within plant cells and how they affect the function of cell organelles such as ribosomes and mitochondria. Phenolic substances that inhibit plant growth (hydroxy derivatives of cinnamic acid, coumarin and naringenin) are synthesized similarly to other phenolics (Fig. **1**). The synthesis of growth inhibitor derivatives of hydroxycinnamic acids follows the pathway: Shikimic acid-chorismic acid-prephenic acid-cinnamic acid and www.biochemj.org/bj/380/0611/bj3800611.htm p-coumaric acid. A theory of metabolism bifurcation among phenolic substances, some of which can inhibit growth and synthesis of indo lic compounds has been proposed. According to this new approach, indole-3-acetic acid (IAA) becomes the main natural auxin [3-5]. Therefore, indole auxins (IAA, indoleacetonitrile) as well as phenolic inhibitors (p-coumaric acid, coumarin, naringenin and others) are derived from the common precursors, shikimic and chorismic acids.

Muzafarov *et al.,* [6] investigated the functions of phenolics in chloroplasts. They assumed that the essence of the relationship between photosynthesis and phenolics biosynthesis is that phenolics exert a direct and an indirect effect on the process of solar accumulation itself. From our point of view, flavonoids as poly functional compounds in green plastids fulfill three major functions as:

- Substrates (use polyphenols and their catabolic products for other kinds of biosynthesis),

- Energy sources (electron and proton transport, ion exchange and membrane potential, radical's formation),

- Regulators (involvement in enzyme reactions as inhibitors or activators).

During photosynthesis under light, flavonoids change the rate of electron transport and photophosphorylation, bringing about the change of ATP/NADPH ratio. In the reactions of carbon metabolism they can shift the dynamic equilibrium of pentosephosphate reduction cycle to enhance the synthesis of certain metabolites both due to the change in energy substrate intake and to the interaction with enzymes of the cycle. Additionally, flavonoids exercise a feedback control over their own biosynthesis, although this phenomenon is not clearly understood. This questionable situation remains as the biosynthesis of the entire flavonoid structure within plastids has not been explained, nor the complete enzymatic package of their biosynthesis has been discovered yet. Lack of direct evidence of flavonoids transport within the cell and through the whole plant constitutes another challenge to a more accurate description of their functions. Nonetheless, a variety of phenolic compounds, present simultaneously within cells appear to be capable of influencing the rate and direction of plants metabolic activities. Thus, any change in the flavonoid structure, or qualitative composition of the phenol complex result in a change of the mechanism of its effect upon the processes of cell energy exchange.

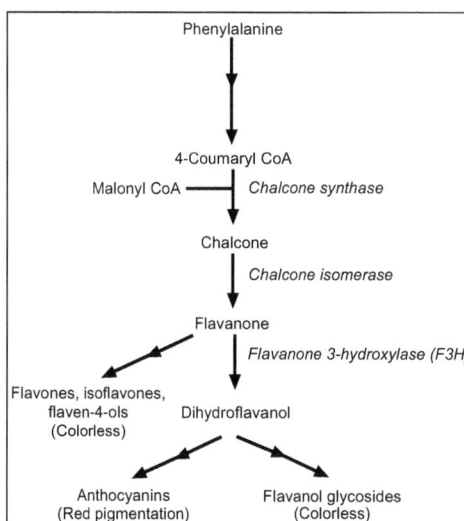

**Figure 1:** Flavonoid biosynthesis (www.biochemj.org/bj/380/0611/bj3800611.htm).

Chalcone and phenol carbonic acids present in etiolated willow shoots can be viewed as the potential precursors of light synthesized flavonoids. However, the use of paper chromatography to investigate isosalidpurposide transformation products did not reveal the presence of any flavonols sensitive to conventional reagents. Therefore, the transformation of chalcone (isosalpurposide) in lightless *in vitro* appears to terminate at a second stage. The synthesis of eriodyctiol and luteolin that occurred in willow leaves evidently took place *in vivo* and under light exposure. It should be pointed out however, that phloridzin and isosalipurposide were decomposed from aglycone and that phloridzin and phloretin produced yellow stains on the chromatogram as well as flavonoids. It is known that flavonoid glycosides are revealed as dark spots on chromatograms exposed to UV light. Therefore, our yellow stains were classified as flavanones, since they did not react with $AlCl_3$, or $Na_2CO_3$, like flavonols that also form yellow spots. At the same time, similar to chalcones and aurones, these floridzin transformation products are yellow colored and they turn into orange-pink when exposed to $Na_2CO_3$ or $NH_4OH$. Relatively easy transformations of isosalipurposide and phloridzin into compounds of other classes (flavanones, chalcones, or aurones) evidenced the role of these products in the general metabolism of flavonoids.

## PHENOLIC SUBSTANCES SECRETED BY ROOTS AND LEACHED FROM LEAVES

Plants contain and secrete a diverse group of growth inhibiting substances that may affect other plants development, if grown in their vicinity (allelopathy). Leaf exudates of willow species such as *Salix rubra* or *Salix viminalis,* contain phenolic inhibitors like naringenin derivative isosalipurposide. Other species instead like apple trees *(Malus spp.)* contain phloridzin, which is a strong respiratory inhibitor. Roots and leaves of the wild plant nanaphyton native to semi-desert regions of Mongolia contain also strong phenolic inhibitors. Seed as well may secrete allelochemicals. Tobacco seed *(Nicotiana tabacum)* for example suppress germination of its own seed when leachates come in contact with the seeds [5]. Although the inhibition of germination was observed at various levels of intensity, this phenomenon demonstrates the selectivity of these natural excreta, similar to the effect of synthetic herbicides. Therefore, increasing evidence indicates that phenolics and alkaloids play the role of selective agents. Secondary compounds can be modified in transgenic plants and genetic mutants. Hence, molecular genetics becomes a tool, which may help to regulate the level of secondary metabolites in plants. Therefore, there is a need continue the search for botanical herbicides as a rise of ecological concerns has clearly identified the environmental impact of herbicides of synthesis.

Root exudates affect the germination of seeds of different crops: monocots and dicots. However, it must be pointed out that only some phenolics were studied in the exudates of willow roots which have no analogues in the roots and leaves found among the common allelopathogens. Although some of these substances could be retained by willow roots, others where excreted into an external medium. Chromatography of these water exudates and a subsequent investigation of their chromatograms with UV-B light showed that most of these substances are polyphenols such as coumarin, or phenolic acids. The phenolic substances retained by cells had different chemical properties than those located in the root exudates. Thus, the data confirm the hypothesis that excreted substances had an allelopathic nature and were involved in developing ecological relationships with adjacent plants of different species (Fig. **2**).

During the composting process water extracts contain many inhibiting substances that might form toxic exudates [7]. Paper chromatography reveals the presence of phenolic acids and coumarins in water extracts. The highest concentrations of these inhibitors were measured in abscised leaves of red maple *(Acer rubrum L.)*. One gram of dry leaves was mixed in 29 ml of water to prepare the extracts. The pH of the solution was between 5.4 and 5.6 and the extracts were incubated for a week at room temperature while the pH rose to 7.2. Further observations revealed that during composting the amount of phenolics was drastically reduced. Seed germination tests were performed with these water extracts and pure water (control) on lettuce and wheat seeds. Germination rate and seedling lengths were measured to demonstrate that phenolics decreased inhibiting properties after dilution.

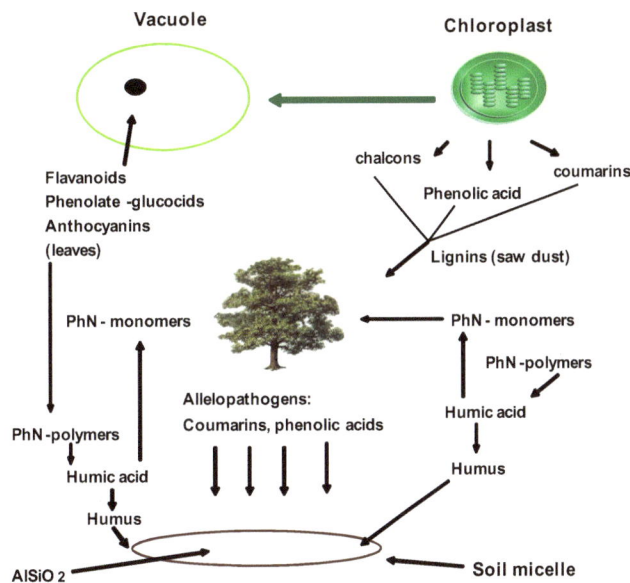

**Figure 2:** Secondary substances, plant biomass accumulation and humus formation during allelopathic effects [8].

## SOIL-MICROBIAL COMPLEX FOR PHENOLIC DECOMPOSITION

Phenolic substances are the most resistant metabolites produced by plants. They undergo further transformation in the soil, forming humus molecules, strongly linked to the alumino-silicate matrix. Humus is more or less a stable fraction of soil organic matter; it adsorbs mineral elements that serve as important nutrients for plant growth and development [9]. The alumino-silicate matrix and humus form primary soil units. Humus is formed by carbon-nitrogen interaction. Potential sources of carbon include cellulose and polyphenols from plant leaves or transformed lignin polymers.

In order to verify the efficacy of microbial activity during the humification process, four different soil horizons in a Gresham soil at the Macoskey Center of Slippery Rock University of Pennsylvania, USA were investigated. The presence and number of colonies of heterotrophic soil microflora were determined in each.

## CONCLUSION

Microorganisms have the capability to decompose phenolic compounds to their monomers, being deglicosidation of phenolic molecules, followed by lignin decomposition, the biochemical pathways of the process. Leaves become a primary substrate for soil microorganisms, while woody materials and sawdust serve as secondary type of biomass and these substrates play a major role in humus formation.

The biosynthesis of phenolic substances within chloroplasts and its further transformation on the alumino-silicate matrix of soil micelles led us to conclude about the existence of phenolics cycle in the plant-soil system. Although many aspects remain unknown, the ecological relevance of phenolic substances in the environment has been amply demonstrated as this cycle embrace lithosphere, microsphere and biosphere.

These emerging concepts facilitate the understanding of complexity within our living systems and their physical habitat while reinforcing the idea of interconnection among living species and ecosystems.

## REFERENCES;

[1]   Kefeli VI, Kalevitch MV. Natural Growth Inhibitors and Phytohormones in Plant and Environment. Kluwer Acad Publ. 2002a; pp. 1-310,

[2]   Lewis NG, Yamamoto E. Lignin: occurrence, biogenesis and biodegradation. Annual Review of Plant Physiology and Plant Molecular Biology 1990; 41:455–496,

[3]   Kefeli VI. Natural Plant Growth Inhibitors and Phytohormones. Dr W Junk b.v. Publishers, The Hague/Boston, 1978, pp. 277.

[4]   Kefeli VI, Dashek WV. Non hormonal stimulators and inhibitors. Biol Rev Cambrige. 1994; 59: 273-288

[5]   Kefeli VI, Kalevitch MY. Natural Growth Inhibitors and Phytohormones in Plant and Environment. Kluwer Acad Pub 2002; 1-310.

[6]   Muzafarov EN, Zolotareva EV. Uncoupling effect of hydrocinnamic acid derivatives in pea chloroplasts. Biochem Physiol Pflanzen. 1989; 184: 363-369.

[7]   Kefeli VI, Borsari B, Welton S. The isolation of inhibiting compunds from the leaves of red maple (*Acer rubrum* L.) for the germination and growth of lettuce seed (*Lactuca sativa* L.). Annual Meeting of ASPP. J. Plant Physiol Abstr 2001; 433.

[8]   Kefeli VI, Kalevitch MV, Borsari B. Phenolic cycle in plants and environment Journal of Cell and Molecular Biology 2003; 2: 13-18.

[9]   Kefeli VI. Fabricated soil for landscape restoration. SME Annual Meeting 2002; Abs Number 142,

# CHAPTER 17

## Chemotaxonomy of Willow Species

**Abstract:** There are some specific substances of the phenolic nature between different willow species which will help to identify the willow species in the comparison of morph- and physiological property of each specie. Among flavonoids, coumarins and phenolic acids there are some specific products corresponding to some species which help to determine the willow systematic property.

## INTRODUCTION

Willows could be good objects for water cleaning and landscape rehabilitation. Their reaction on auxin and phenolic quality could be used as chemo-taxonomic factors. In comparison with morphological properties, they could be applied for willow species identification [1].

Phenolics are the secondary products of photosynthesis which participate in xylemogenesis (lignin formation) or are deposited in the form of glycosides in the vacuoles. They are easily observed on paper chromatograms in UV-B light. There are many functions of phenolics in plants. They participate in respiration as substrates for oxidation. Some of them modify auxin activity in the plants, and some are substrates secreted from the roots (inhibiting and sanitary properties). The quality of phenolics may help to identify different plant species, as it was shown by Harborne [2].

Willow cuttings were rooted in water or in indole-acetic acid, 150 mg/l. The type of rooting might be a factor of chemotaxonomy.

Leaves of different species of willow growing on FS [3, 4]     were extracted by water in a ratio of 1:5. Water extract was applied to Whatmann I paper chromatography using 5% acetic acid as a solvent, ascending way. The chromatograms were dried, treated by saturated soda solution and investigated in UV-B light.

Some physiological and biochemical properties of willows were used for willow species determination. The most pronounced physiological process is rooting and the effect of auxins on it.

The data show that different willow species have different properties of rooting in the fall and different reactions on auxin. These parameters could be used as a chemo-taxonomic factor.

The other factor of chemo-taxonomy is phenolics which were observed on the chromatograms as different fluorescent spots. Flavonoids and their glucosides were identified using Rf, color reactions, and change of color by soda reagent. The same parameters were used for the identification of coumarins and phenolic acids [1].

Two types of flavonoids were observed for willows, one with a low Rf and the second with a higher Rf. They are easily separated from each other. Silky and pussy willows contain only one flavonoid with a low Rf. Silky willow also has a low Rf for coumarin. Caffeic acid was present in all willow species and therefore could not be used for willow taxonomy.

For chemotaxonomy of willow species, the investigation of flavonoids and coumarins could be recommended. Phenolic acids are not specific for the process of identification. Chemo-taxonomy might be a supplemental tool for identification of willow species. It might also be very important for the investigation of the cross-pollinated forms.

## SUMMARY

The reaction of cuttings of willow species on auxin could be a factor for chemotaxonomy. The pattern of phenolic substances in willow leaves was also different for the six investigated willow species [1]. Among the phenolics are flavonoids of the salipurposide and luteolin family, chlorogenic acid, and oxi-coumarins. Leaves of the six willow species were investigated on the qualitative characteristics of observed phenolics in the six willow species. Water extracts of the leaves were subjected to ascending paper chromatography (solvent 5% acetic acid). Willow flavonoids (Rf 0.1-0.4) were identified by Rf, UV-B, soda reagent, and soda + UV-B. Both silky and pussy willows had one flavonoid and weeping willow had two flavonoids. Autumn, purple and white willows had 3-4 flavonoids (luteolin, salipurposide, isosalipurposide). All flavonoids are in glucoside forms (Table **1**).

The flavonoid complex was isolated for each willow from Whatmann 3M paper and rechromatographed in 70% isopropanol for further identification. Water soluble extracts from the leaves were applied to silky willow cuttings before rooting. Extracts from leaves of purple and white willow (deep dormancy) inhibited the rooting process. Willow flavonoids could be used as tests for willow chemotaxonomy [2].

**Table 1:** Comparative data on the willow species leaves water extracts with ascending paper chromatography.

| | Rf | | | |
|---|---|---|---|---|
| **Species** | **Flavanoid I** | **Flavanoid II** | **Coumarins** | **Caffeic acid** |
| Silky | 0.06 | | 0.6 | 0.78 |
| Pussy | 0.05 | | 0.48 | 0.68 |
| Weeping | 0.06 | 0.33 | 0.64 | 0.88 |
| White | 0.06 | 0.23 | 0.61 | 0.73 |
| Red | 0.06 | 0.17 | 0.46 | 0.67 |
| Autumn | 0.06 | 0.3 | 0.5 | 0.68 |

## REFERENCES

[1]    Kefeli V, Kalevitch MV. Natural Growth Inhibitors and Phytohormones in Plants and Environment. Kluwer Academic publishers, Dorderecht/Boston/London, 2003. pp. 322.

[2]    Harborne J. Phenolic Glucosides in: The Biochemistry of Phenolic Compounds. Academic Press, NY. 1964; pp.129.

[3]    Kefeli V. Water Cleaning and Wetland Construction. International Journal of Environment and Pollution 2007; 29: 383-391.

[4]    Kefeli V, Dunn MH, Johnson D, Taylor W. Fabricated soils for landscape restoration: An example for scientific contribution by public-private partnership effort. International Journal Environment and Pollution 2007; 29: 405-411.

# CHAPTER 18

# Water Cycling and Plant Management

**Abstract:** Plants are considered as biological filter for waste water cleaning and reuse it. The population of papyrus plants is applied for gray water transformation. Black water and humanure compost could be used for perennial plants, trees and shrub propagation. These cycles could be applied for closed eco-systems creations.

Waste water and humanure were investigated for their ability for use as substrates for plant growth and crop yield production for research and potential application to the Building as Power Plant (BAPP) project at Carnegie Mellon University (CMU). Biological activity of wastewater (gray and black) and composted humanure was determined using tests on four plants: wheat, clover, mustard, and lettuce. Gray water and humanure did not possess any toxicity, but black water was found to be toxic and to inhibit seed germination and seedling growth. Humanure used as a compost component in the FS (FS composition) for the growth of tomato and cucumber seedlings was similar to that of the control (topsoil or standard FS). However, harvest results of cucumbers and tomatoes demonstrate that the application of com posted humanure enhanced the cucumber and tomato harvests in comparison to the control. Recycled waste products, water and humanure, could be incorporated into indoor vegetable and plant production. It is recommended that this research be utilized by planners of the Building as Power Plant (BAPP) project at CMU for incorporation into the waste cycle.

## INTRODUCTION

This research serves as a companion report to Sustainable Water Management for the Building as Power Plant (BAPP) at Carnegie Mellon University report from spring of 2005. The previous report focused on potential water management strategies for incorporation into the design of BAPP. This report reflects on investigations regarding the potential utilization of waste products (wastewater and humanure) within the design of BAPP. These waste products could be used throughout BAPP to close the waste cycle and provide for a self-sustaining system.

### INTEGRATION OF ARCHITECTURAL CONSTRUCTIONS WITH PLANT RESEARCH

- Propose a mixture of waterless urinals, low-flush toilets and waterless, composting toilets, such as the Clivus Multrum (CM) [1].

- Estimate the amounts of "gray" water and "black" water that result from specific distributions and various usage patterns (i.e. during the semester, as well as off-peak times).

- Identity the biological processes and associated plant and nutritive materials, such as saw-dust, wood-chips, and leaves etc., needed for the formation of useful FS.

- Identity the amounts and predicted characteristics of various types of FS to associate with potential uses. This applies to growing food (tomatoes, cucumbers) or cultivating decorative plants. This study also includes the impact of plants on indoor air quality, as well as soils for living machine applications.

- To coincide with our biologically focused studies, preliminary designs are proposed for the configuration and layout of dedicated spaces within the BAPP/Invention Works project. These spaces will include greenhouses on the ground, greenhouses integrated on the top floor that double as livable spaces, as well as atriums and green roof technology.

### INVESTIGATION AND RESULTS

1. Proposal for mixture of waterless urinals, low-flush toilets and waterless, composting toilets, such as the Clivus Multrum (CM).

The utilization of waterless urinals, low-flush toilets and composting toilets provides for an internal biological cycle working within the BAPP. Waste material can be used as nutritive sources of nitrogen, phosphorus and potassium.

Additionally, waste material can be utilized in a FS composition for decorative plants and vegetables. An experiment was initiated using CM humanure (H) waste as 1/6 part of the FS mixture. The controls for the experiment were top soil and regular FS.

2. Estimation of the amounts of "gray" water and "black" water to result from certain distributions and various usage patterns

While specific usage patterns won't materialize until the barn design is more developed, research data on gray and black water may assist in planning. Gray water is suitable for application to most plants, while black water must be diluted heavily before reaching plants. Research uncovers that it is recommended for black water to be diluted with 20 parts water to provide for adequate plant germination and growth.

3. Identification of the different kinds of biological processes and associated plant and nutritive materials that would be needed for the formation of useful FS.

All components of FS maintain their unique role in the biological processes. The following are components of the FS recipe known as LL or living lab recipe:

- Saw dust provides a long-term source of carbon

- Dry leaves provide a short-term source of carbon

- Mushroom compost provides source of nitrogen

- Humanure provides an additional source of nitrogen

- Pond sediments provide a source of clay for water retaining

- Coffee residues provide both carbon and nitrogen

Preliminary testing was conducted on all components of the mixture to identify the presence of chemical and physiological properties, bacteriological flora, and general fertility. Research is on-going in order to refine the recipe to provide the most productive mixture for outdoor plants (established in ponds and irrigation fields), indoor plants (for cloning and enhancement of the indoor wetland and sanitary plants), and plantlets for greenhouse and seed beds.

4. Identification of the amounts and predicted characteristics of various types of FS to associate with potential uses. This applies to food growth (tomatoes, cucumbers) or decorative plant cultivation. This study also includes the impact of plants on indoor air quality, as well as soils for living machine applications

The LL FS was tested for the growth of tomatoes and cucumbers. This soil was compared to the control, standard FS and topsoil. Research results are presented later in the paper. Growing plants indoors provides clean air through volatile essential oils and supplies a significant amount of oxygen to the indoor air composition. Additionally, the introduction of plants indoors becomes part of the phyto-design and the plants grow without additional light under low energy conditions.

5. To coincide with our biologically focused studies, we propose preliminary designs for the configuration and layout of dedicated spaces within the BAPP/Invention Works project. These spaces will include greenhouses on the ground, greenhouses integrated on the top floor that double as livable spaces, as well as atriums and green roof technology.

Preliminary investigations were made in the IW for the use of modeling atriums with decorative plants in the office environment. Plants were developed utilizing FS and maintained through the study. BAPP has the potential to build upon the initial investigations of indoor plants and implement multiple greenhouse spaces throughout the structure. These spaces will provide many benefits to the building, including improved air quality, temperature moderation and various aesthetic benefits.

Both substrates, FS and H, were suitable for cucumber production. However, humanure forced more green mass production (stem+leaves) than harvest (cucumber fruits). All substrates: topsoil, FS, and FS+H, were proper for tomato fruit production. The best compositions of the soil for fruit productions were top soil and humanure while FS was less active. The soil mixture of humanure with the other components of the FS is a good substrate for vegetable (cucumbers and tomatoes) production. No *E. coli* were observed in both soil mixtures (US Microsolutions Inc.). The plants could be cultivated in the form of container gardening and could be the element of waste cycling in the system of urban agriculture.

# REFERENCE

[1]    Jenkins JE. The Humanure Handbook: A guide to composting human manure. Jenkins Publishing 1999; pp.301.

# CHAPTER 19

## Ecological Modeling in Space

**Abstract:** Conception of biological processes in Space is also based on plant – soil-water relations in closed systems. The ideas of reuse of waste water in plant growth and application of some plants as biological filters could be applied in the control of biological processes in the space. Relations of plants with microorganisms and soil components in the space, in the closed eco-models are considered in this chapter. It is important to consider the effect of solar energy for these biological models, taking into consideration the processes of photosynthesis, photomorphogenesis and photoperiodism.

Previous research described in the special chapters could be organized in the chain of reactions which could be applied not only on the earth but also in space. Consider plants and algae as oxygen producers (phytosphere) [1]. Plants as autotrophy systems could be the fragments of the complex artificial eco-systems with dense canopy, resistant to loading, positive reactions on low dosages of fertilizers N, P, K. At the same time, the artificial soil (FS) could be a center of the activation of symbionts and free microorganisms, which activate composting processes, humus formation and improvement of soil aeration. All these components of plant - soil relations could form the conception of space agriculture. Of course, there were many specific properties of the space agriculture. Recent publications about space agriculture indicate that a major objective is to pursue research not possible on earth, by escaping the familiar effects of gravity and entering an environment known as microgravity. It is clear that cells behave differently in space, which might be based on the differences in fluid flows in microgravity. Space agriculture will be critical to future space travelers. Scientists view plants as a way to explore the behavior of cells in microgravity. How plants respond to microgravity is the focus of investigations for many scientists. This research will let us ultimately learn what controls cell division and the inheritance of genetic information.

## INTRODUCTION

Transition of elements from one block to another is a significant factor in integration of biological units into the ecological system. These blocks form a certain chain or a cycle where interactions between blocks play the most important role. For example, a block of plants has several outputs. They include organic matter formation for heterotrophic organisms and autotrophs, or excretion of volatile products into the surrounding atmosphere. These outputs are symbolic "gates" through which plant metabolites can move in both directions in or out the plant community.

The function of metabolites as nutritive, stimulating, or inhibiting agents in the interrelations with other block systems is important. The example of such interactions might be either microbiocenoses or animal/human community. The stability of the biological block usually depends on its internal life plus linkages and interactions between blocks. Thus the stability of a biological system in space depends on the block stability and on the input of each block to the entire ecosystem. The life of biological blocks depends on the life within cenoses based on density tolerance, allelopathic regularities, and ecological factors. Hence, we need to elaborate on the system of blocks. The specific project will be the evaluation of the processes occurring in each block individually. However, the regulating actions between blocks are significant because they stabilize the activities of each block whether separate or combined.

Green plants in space serve as a source of:

1.  Gases, which maintain stable atmospheric conditions,

2.  Organic matter, and serve as a source of nutrition for humans in space and as a component of the food chain,

3.  Organic matter for the growth and development of heterotrophic organisms such as fungi and bacteria.

Based on this approach, green plants should be divided into two main groups:

1. Flowering autotrophs with stems and roots,

2. Non-flowering autotrophs such as algae and bacteria.

The main purpose of this project is the establishment of the cyclic independent systems that will function for a long term in space. The following problems which enable us to establish ecological models in space must be solved:

- the orientation of plants under space gravity (microgravity)

- the re-utilization of organic substances secreted by plants, human, and algae

- the application of lower organisms as a dependable source of enzymes, which split excretions and accumulate essential growth stimulating substances

- interaction of green plants in the crop rotation systems, including the investigation of allelopathic reactions;

- reaction of living organisms on stress conditions (UV-light, $CO_2$-deficiency, low and high temperatures).

The whole conception of "Ecological models in space" should be presented in at least three main blocks:

1. Biotechnological models in space;

2. Plant physiology and biochemistry as supplementary tools for ecology in space;

3. Principles of agronomic techniques as support for space ecology.

## BIOTECHNOLOGICAL MODELS IN SPACE

The specific feature of above mentioned combination is the construction of stable biotechnological blocks with advanced components [2]. Under these circumstances plants ontogeny is coordinated with life cycle of microorganisms to form integrative biotechnical units. The biotechnological concept allows us to construct these special blocks for plants cultivation and for successful growth of microorganisms. A reservoir, where absorption is occurring, is a significant element of the system, because excretion products (EP) both from plants and microorganisms must accumulate there. The biological activity of these EP will be analyzed [3].

It is significant to isolate EP from biotechnological system to determine if the concentration is high or toxic or if they are able to suppress the further development of culture-producer or partner culture. When the situation is positive and the excreted products are stimulators, the media will be enriched with nutrients. These nutrients will return to the cultivation block or to the tank with other culture. For example, some bacteria and algae produce phytohormones which activate plant growth. On the other hand, plant excretions are essential for microbiotic growth. Hence, in general non-flowering organisms (algae, fungi, and bacteria) are the source of growth substances for successful development of green plants.

Non-flowering organisms can produce phytohormones (growth substances) which:

- Stimulate plant growth reactions (for example gibberellins produced by fungi *Fusarium moniliforme).*

- Form roots or stems (for example auxins which are biological products of *Agrobacterium rhizogenes* or *Taphrina sadebekii*)

- Induce the multiplication of cells in tissue cultures or in green water plants such as *Sphyrodella olearizae* (for example cytokinins produced by fungi *Botrytis cinerea*).

Green mass of plants, even plant residues can be a potential source of organic substances which further can be used for the growth and development of above mentioned microbiological cultures. Thus, lower organisms are important components of ecological systems in space. They can successfully utilize various organic substances for their growth and development.

This study will be based on the obtaining of active races of microorganisms which serve as stable producers of growth substances and active enzymic forms. The problem of submerged types of bacteria and fungi cultivation is important.

## GREEN FLOWERING PLANTS AND GREEN ALGAE

Determination of plant density tolerance is one of the primary conditions which make plant cultivation successful in the block of artificially designed community. Plants of the same species, but different varieties, demonstrate different reactions under various density conditions. It is significant to select the proper plant variety for further cultivation in space considering that some plants absorb much more food than others and are more sensitive to light intensity. Similar selection must be performed for green algae. It is important to select flowering plants according to their geo-orientation.

In space green plants usually lose their normal orientation because gravity fields are absent. Therefore one of the main purposes of this study is to search for a remedy to plant axal systems orientation with respect to water, nutrition and light which cause topic irritation. The possible remedy is the construction of a cassette system where limited orientation of seedling's stems and roots is present.

The reaction of green plants to light quality, intensity, and period is a significant parameter for plant growth and development. However, the propagation of flowering plants is also an important element of plant's life. Therefore, hormonal substances, produced by fungi, algae, and bacteria are an essential contribution for successful performance of this project.

The production of fertile plants in space is a more complicated process compared with propagation of flowering plants. The short-term vegetated plants like *Arabidopsis* or *Chenopodium* are not efficient for this project, because they don't support some practical aspects of this research. Therefore, it is important to find such types of plants which can easily form seeds in space (this will be a special study). The production of biomass in space is a significant matter.

One of the most important aspects of this project is to select crops which propagate easily, and are characterized by:

- very intensive biomass production;

- complete utilization of plant for feeding purposes;

- substrate and source of nutrients for lower plants propagation;

- sufficient chemical content of green food, including proteins (essential amino acids), carbohydrates (monosaccharides) and lipids.

Examples of such quick growing cultures might be several varieties of *Amaranthus*, as well as some lettuce and cabbage cultivars. Thus biotechnological models for space technology include higher plants, several water plants, algae, bacteria, and fungi.

So, food menu from green mass and microorganisms have to be a result of combined growth of these objects and cooperative utilization of their excretions, residues, and hormonal products which will be further used for stimulating processes in system.

The idea to utilize waste residues for further cultivation of other biological organisms is an example of interrelations within the biological chain.

It is essential to construct special equipment such as cassettes and chambers (they will work in joined chain) for successful cultivation of biotechnological models in ·space. The components of this complex are: the chamber of plants disintegration, mycelium and algae cells residues or elementary components which must be further reutilized in the nutrition processes or proceed for further disintegration of components till mineralization. Some immobilized forms of enzymes could be used under these circumstances as the factors of quick organic matter disintegration.

Sometimes these substances are unable to eliminate the effect of inhibitors which were excreted by the other plants while their growth and development. Hence, we should assume the sanitary properties of natural soil which absorb and destroy the effect of inhibitors. The level of natural growth inhibitors, which work as allelopathogens, and were excreted by roots of cultivated plants, can be managed by growing these plants in cassette forms. This control will permit us to reutilize substrates several times if the amount of inhibitors after plant growth is minimal in the system. Sometimes it may be necessary to develop the enzymic complex which will completely split the inhibitory (allelopathic) complex. It is significant to use active forms of microorganisms which can split the allelopathic inhibitors.

Cultivating plants on the same media for a long period of time without changing growth activity of plants is a complicated process. However, there are distinct examples of long term cultivation on the same place without further rotation. Rice in a paddy form exist symbiotically with fern (green algae complex), and is a good example. Fern fixes nitrogen for higher plants, and also serves as a good sanitary object.

Some genetic performances are not excluded for successful growth of plant-microorganism complex in space. So, the symbiosis of plants and algae or heterotroph-microbial complex performance in space should provide a proper gaseous atmosphere, essential food menu, and influence positively on plants growth and development.

The higher plants block is tightly connected with microbiological and algae excretions and their reactions. The stability of this complex depends on the common exchange reactions, and it is certain that physiological and biochemical processes play an important role in this system.

## PLANT PHYSIOLOGY AND BIOCHEMISTRY FOR ECOLOGY IN SPACE

Plants and cultures in space are developed under such conditions which could be easily determined as stresses: microgravity, disturbance of photo- and thermoreactions, oxygen deficiency, etc. Plants and microorganisms which we plan to use in space modeling must be preliminary evaluated on earth because of their reactions under stress. The most resistant forms must be selected, and then utilized in space. The affect of two or more stresses in combination will be studied. One of those stresses might be a factor which increases plant resistance. Some inert substances like ceramsite or vermiculite may substitute for soil as a source of mineral elements for plant nutrition.

The ultimate aim of the study is to cultivate ecologically pure material in space considering that leaf and root types of vegetables can grow rather fast. One of the ways to achieve this goal is to induce plants growth under gravity, and later expose plants to space microgravity using strong orientation factors like nutritional (chemotropism) factors or water (hydrotropism) regime. Light conditions and density tolerance are also important factors of plants growth. Some plants can successfully survive under high density of sowing, and form green mass better than others. Therefore it is significant to provide preliminary screening of the plants, and evaluate their reaction to common stresses. It is also essential to check the ability of plants to synthesize nutritive substances like proteins, carbohydrates, and lipids (quality products). The ability of plants and algae to maintain stable gaseous atmosphere is also an important issue in space biology. Thus biochemical composition of each product excreted by plant should be thoroughly investigated.

We would like again to stress the importance of density tolerance in space, which leads to the selection of plants with special features. These plants demand less nutrition, so density tolerance is such a characteristic which allows plant to survive and vigorously develop in a plant community, or agrocenoses. The shape of the plant which depends on its genic features is also an important parameter in artificially designed cenoses. Physiological processes which take place during plant ontogenesis and depend on such basic parameters as growth, productivity, resistance to unfavorable factors, root nutrition, and photosynthesis are considered to be rather significant in space design. Some of these parameters can be easily modified as viewed from genome and hormone regulation, but other features, for instance, photosynthesis, needs more complicated remedy to be modified for space study.

Thus in any case, the ultimate yields depend on the increase of biomass as well as on the transformation of photosynthate from the leaf to other vitally important parts of the plant, i.e. root, fruit, tuber, and ear. The increase of the biomass is the result of cooperative interaction of two processes: growth and photosynthesis. However, the leaf metabolites do not accumulate in the vital parts of plant organism without functioning of transport systems, hormone regulators, and well-arranged donor-acceptor relationships. So, plant productivity is a phenomenon which basically is the mutual interdependence of the main physiological processes occuring in plants. The understanding and description of these processes reveal the role of defoliation in crop formation. The leaf is the center, where primary products of photosynthesis are forming. The leaf is an organ where metabolism of these products occurs. These products later evacuate into reserve organs, with further subsequent aging of the leaf itself.

The primary mechanisms of ontogenesis include formation of morphological structures, function of meristem, formation of leaf primordia, and correlations in the entire plant. All these processes are controlled by genome and by the system of phytohormones. Each organ, including leaf, experiences with the whole plant the main stages of its development: embryo formation, juvenile stage, maturity, and senescence. The final stage death is followed by the removing of the plant organ from the place of its dwelling. It is doubtless that this interconnection serves as the basis for the leaf development which is formed by the integral plant organism.

The leaf as the center of assimilates formation earns special attention because of plants productivity. The important element of leaf vital activity is a stage of its growth accompanied by the development of its photosynthetic functions. The ability of the leaf to donate these assimilates, as well as to evacuate these metabolites sideways through the growth point of stem or root, are also a significant characteristic of the leaf functioning. In this case the peculiar cascade of "transfer stations" is formed. These stations are considered to be a system of independent pairs: donors and acceptors. The pair leaf parenchyma -phloem ends is localized at the beginning of this system. The pair leaf cutting- stem phloem is the next one, and etc. In each of the given pairs the hormone gradients play a tremendous role because these gradients serve as channels by which metabolites are transported from the leaf to the vital organ.

Auxins and cytokinins are the most active hormonal metabolites which are involved in the formation of sink (acceptor) center. The role of abscisic acid in evacuation of metabolites is not completely clear.

The defoliation itself is closely connected with crop formation and productivity. The leaf as a source of metabolites in the final stage of ontogenesis definitely proves this becomes a donor of metabolites which further are transported into different parts of the plant. Simultaneously with the transport of assimilates leaf aging occurs, which is the final stage of ontogenesis, followed by defoliation. Meanwhile, the principle of "export integrity" of metabolites is essential. Some low-productive varieties are characterized by low acceptor rate or even the complete absence of this activity. So the metabolites are transported to the axillary vegetative organs. The described above variety specifics complicate the defoliation. Chemical defoliants are used to correct these natural mistakes or mistakes of breeding, and facilitate the process of defoliation.

Photosynthesis and plant growth should be considered with plant soil nutrition. The interaction of soil and air nutrition can be viewed by cooperative investigations of specialists on the system analysis. The carbon cycles in photosynthesis as well as the nitrogen cycles are the basis of productivity.

The analyses of the unfavorable tendencies in plant growth readily demonstrate their magnitude and long-term character. Therefore, inevitability of marking new priorities of plant growth intensification provides new stage of its development in the interest of the human community.

The optional significance of contributions of non-compensated energy (from fertilizers, pesticides, and irrigation, etc.) is revealed, considering that the basic role of photosynthesis is crop formation and soil fertility. The share of this energy even in the most technologically intensive agrocenoses relative to the entire energy of the sun ranges about 0.05%. However, the real importance of the stream of non-compensated energy should be estimated not only from the point of view of special relations but from the standpoint of quality functions. The non-compensated energy should be regarded as an important mean for controlling the large streams of energy in agrobiocenoses.

So, it is important to study the growth mechanisms of plants in space in consideration with photosynthesis, plant metabolism, and the ways of plant adaptive systems performance in agricultural production. The search of information indexes the dynamic model and its physiological characteristics of new non-traditional ways to control the productivity and resistance of agrocenoses in space. The stability of agroecosystems depends greatly on biological structure of agrophytocenoses including resistant plant components: mutants, transgene plants of allocytoplasmic hybrids.

The outstanding achievements in plant physiology at the beginning of 20th century, opened the new era of the "green" revolution. However, now scientists realize that the processes of intensification of agricultural production, its large-scale concentration and specialization cause a number of new fundamental scientific, social, and ecological problems. They comprise the growing value of each additional food calorie, the increasing emergency of disturbance of ecological balance and environmental pollution. The requirement is to protect agriculture more intensively from the uncertainty of nature. The most acute problems stem from utilization of artificial and natural energy resources on the agroclimatic arrangement of the cultivated plants, as well as development and application of energy-saving technologies. Thus, the increase of adaptive potential of the agricultural productive system leads to the increase of general and specific adaptability of cultivated plants.

The quality of green mass and yield of seeds and fruits is also rather important, because it will serve as a food source in space. The quality of the plant material depends on the genic properties of the plant, on its biochemical components (sugars, amino acids, and lipids) which control the composition of these particular substances in plant. Some plants are unable to produce the essential amount of amino acids or sugars. In this case it is significant to compile a special "menu" which supports plants, like utilization of algae, bacteria, or fungi excretions. This approach will provide humans with essential nutrients in space. It is also important to consider that during long term flight the composition of plants may be changed. Some plant complexes (blocks) may be substituted with others. One should also consider that high temperature or some conservation processes may cause definite biochemical conversion inside plant material.

But beneficial physiological and biochemical features of plants help to sustain proper gaseous atmosphere (evolution of oxygen), assist in providing ecologically reasonable behavior of plants in the community, and obtaining ecologically pure products in space. However, the principles of agronomy seem to be priority importance in space.

## PRINCIPLES OF AGRONOMY FOR SPACE ECOLOGY

The concept of cleaning the substrate from allelopathic substances is performed by rotation of the cultures and use of sanitary active microorganisms.

The most important principle of agronomy which must be used in space ecology is the maintenance of general cycles, especially carbon and nitrogen cycles. These cycles which include P and S will fulfill the turnover of substances in the space and establish living processes for autotrophs and heterotrophs. Some specifics of space biology are in the performance of this idea in a more rapid manner, including selected organisms which might establish biological equilibrium in space.

The most important aspect is the input of each biogeochemical cycle into the general life process. Thus C-cycle is based on the process of photosynthesis (photosensibilization), which is tightly connected with chlorophyll activity. The act of photoreception is very short, its duration lasts pico-second period. The further reaction could be much longer especially in the region of biosynthesis of primary and by-products of photosynthesis.

The longest are processes of organic matter destruction in soil and formation of humus molecules, which are involved in soil fertility management. In general C-cycle is a source of autotrophic processes which generate energy and substances for different biological cycles. The interaction between blocks is occurring due the excretion of metabolites into external media. Green plants as autotrophs and photosynthetic organisms will definitely play a domination role in these interactions. It is important to consider that some organisms demonstrate a certain competition for mineral or organic substrates. The participation of autotrophs and heterotrophs are based on cycle interactions.

The C-cycle is unable to function without nitrogen turnover based on nitrate reduction and $N_2$-fixation. N-cycle cannot function without donation of phospho-organics (phospho-sugars, ATP, phospho-inosite) for the bulk reactions of amino acids, polypeptides and nrntpin linkage between these two cycles. P-containing salts in soluble forms penetrate via root system to the center of photosynthesis or protein biosynthesis. In lower plants this process occurs via cell membranes. The balance between C, N, P, S cycles is a base of normal life processes.

Therefore the selection of organisms for space research must be provided in a special way. These organisms must maintain the cycling processes in a properly coordinated manner. Such a reaction as the turnover between cycles is based on the idea of crop rotation combined with the properly designed microbiocenoses. The communities of higher plants which might be used in the system of crop rotation must be accompanied by the constructed microbiological community.

Hence, this project considers complex interaction between all components: plants, algae, and microorganisms by excretion metabolites in volatile and liquid forms. Plants in space are definitely developing under stress conditions, one of which is microgravity. Therefore before the construction of the community it is important to select genetically appropriate partners. For example, plants must not be sensitive to gravity.

The investigation of some physiological and biochemical characteristics is also significant in space biology but the most essential parameter in· this project should be the study of interactions between components. In this case the properties of higher plants and microorganisms must be studied holistically in cenoses. The main aspects of this project should be definitely investigated at first in the laboratory on earth, and later transferred into space conditions of micro gravity.

## REFERENCES

[1]     Kalevitch MV, Kefeli VI, Johnson D, Taylor W. Plant biodiversity in the fabricated soil experiment. Journal of Sustainable Agriculture 2006; 29: 101-114.
[2]     Kalevitch MV, Kefeli VI, Borsari B. Bacterial activity in fabricated soils. Presentation at 104th American Society of Microbiology General Meeting. New Orleans, 2004a, Page number 1.
[3]     Kalevitch MV, Kefeli VI, Borsari B, Davis J, Boulos G. Chemical signalling during organism's growth and development. Journal of Cell and Molecular Biology 2004b; 3: 95-102.

## CHAPTER 20

# Soils for Sustaining Global Food Production

**Abstract:** This chapter summarizes the soil functions for human civilization. The most important is the way of the soil preservation. Land quality is considered from six point of view including agriculture, forestry, human well being, sustainability of eco-systems. The role of soil in the process of the sustainability of human civilization is a main goal of this chapter. Thus, soil is a leading component of bio- sphere, where plants are the direct link between solar energy (photosynthesis) and humus production in the soil.

## INTRODUCTION

The social, economic, and environmental well-being of humans is strongly linked to soil quality. Sustainable development as a goal for all nations depends on the interdependence of these factors and how human society manages natural resources in a harmonious manner. Soil resources have always been important from the time humans ventured into sedentary agriculture. However, during each stage of the development of societies, the issues, concerns, and societal commitment to manage the resource varied. As populations increased with a concomitant increase in the demand for food and fiber, greater effort was made to understand and enhance the performance of soil resources. Modern soil science is only about 100 years old, but in partnership with other disciplines, it has made major contributions to global food security. In the United States, Europe, and other advanced countries, yields of crops have more than tripled in the last 50 years; even in many developing countries the Green Revolution has brought tangible changes. This progress significantly reduced the probability of global hunger and famine. Today, the total food production is adequate to feed the world. Inadequate or inefficient distribution of the food, however, prevents many from reaping this abundance and leading to an estimated more than 800 million malnourished people. While significant advances are being made to enhance the productivity of soil resources, in some countries of the world the ability to sustain this productivity level is being reduced by overexploitation or inappropriate use of the soil resources.

The issue facing national policy makers in most countries of the world is the ability of the land to produce the food and fiber for the growing populations [1]. The issue stems from four fundamental concerns: The first is land degradation, which results in the decline of the quality and quantity of land. The second is population growth that constantly threatens the ability of the country to feed and clothe the population. The third is unequal access to resources wherein the affluent have disproportionately more land, forcing the poor to exploit fragile ecosystems and thereby accelerating land degradation. Finally, the fourth is resource consumption whereby land is permanently taken out of agriculture for urban and other permanent structures. These concerns challenge food security, which is defined as "access by all people at all times to sufficient food for an active and healthy life." The net consequence is that environmental degradation, accentuated by human mismanagement of land, is negatively impacting the basic life support system of planet earth and some view this as leading to national, regional, and even international conflicts.

The United Nations Conference on Environment and Development at Rio de Janeiro in 1992 has spawned a number of global conventions to protect and conserve the earth's resources and environment. The shared commitment to enhance the quality of human life while maintaining a balance with the other components of the environment is the goal of all nations. However, the absence of tangible commitments by many nations due to the inability to provide the required investments has yet to result in any meaningful impact. In many developing countries, national policies are either absent or, even in some cases, inadvertently aggravate the problems. The global dialogue has, however, resulted in creating the necessary awareness among nations and international collaboration is being initiated to share the responsibilities.

At a conference on sustainable agriculture held in Italy, the participants posed several questions and concluded that there are numerous alternatives to mainstream agricultural research and development and that their application may determine whether the people of this world can successfully meet their needs. There are many obstacles to this optimistic view, the most important of which is the socioeconomic and political state of developing countries where such advances must also take place and where major obstacles exist as illustrated by Cleaver and Schreiber [2]. This is beyond the scope of the present paper, which will only consider the biophysical aspects and specifically the quality of the land and the forces that detract from enhancing productivity. The issue of food security is further complicated in developing countries when it has been achieved at the expense of the integrity of the environment [3]

Eswaran *et al.,* [4] showed that of the 130.8 million km$^2$ of ice-free land about 14.5 million km$^2$ or 11.1% is arable and used for agriculture and/ or grazing. An additional 2.4 million km$^2$ of land, largely in the arid parts of the world, is irrigated. This 16.9 million km$^2$ of land currently feeds 6.4 billion people, which is expected to increase to more than 10 billion in the next 25 years. Over the past 40 years, per capita world food production has grown by 25%. Yet the world still faces a fundamental food security challenge, with some 800 million people hungry. What is going to be important is who produces the food, has access to the technology and knowledge to produce it, and has the purchasing power to acquire it. Many countries have reached or are reaching the limits of their land resources. Land degradation and other land consumption processes such as urbanization and infrastructure development are continuously reducing the amount of land for food and fiber production.

In this report on global land resources, we first evaluate the availability of land and the condition of the resource base in the context of the functions that have to be performed. In many countries of the world, there appears to be a systematic decline of the quality of the land, which impacts the abilities of the country to be sustainable. In our assessment of the ability of soils to sustain global food production, answers to some of the following questions are provided:

1.    What is the capacity of global soil resources to produce food?

2.    As not every soil has the same capacity, where are the best areas and the hotspots;

3.    How can the comparative advantages be exploited in a national, regional, and a global manner;

4.    What are the needs to feed 10 billion people?

## GLOBAL SOIL RESOURCES:THEIR QUALITY AND DISTRIBUTION

An assessment of global land resources was made by the United States Department of Agriculture [5] and more recent and updated data are presented in Table 1. The global soil map is reproduced in soil taxonomy [6]. In Table 1, the land area occupied by each soil order is given and, in addition, an estimate of the number of people living on such soils is also presented. This is obtained by overlaying a population density map developed by Tobler *et al.,* [7] on the soil map. Ultisols, Alfisols, Inceptisols, and Entisols have the high populations and together support over 70% of the world population. This group of soils occupies about 44% of the land area but members in this group also present favorable conditions for agriculture.

**Table 1:** Global soil and land quality classes.

| Soil and Land Quality Classes | Land | | 2002 Population | |
|---|---|---|---|---|
| | Area (mill sq, km) | % | Area | % |
| 1. Total ice-free land | 130.8 | 100 | 6400 | 99.9 |
| 2. Kinds of soils | | | | |
| Gelisols | 11.26 | 8.61 | 25 | 0.4 |
| Histosols | 1.53 | 1.17 | 31 | 0.5 |
| Spodosols | 3.35 | 2.56 | 107 | 1.7 |
| Andisols | 0.912 | 0.70 | 110 | 1.7 |
| Oxisols | 9.81 | 7.50 | 252 | 3.9 |
| Vertisols | 3.16 | 2.42 | 356 | 5.6 |
| Aridisols | 15.7 | 12.00 | 353 | 5.5 |
| Ultisols | 11.05 | 8.45 | 1148 | 17.9 |
| Mollisols | 9.01 | 6.89 | 428 | 6.7 |
| Alfisols | 12.62 | 9.65 | 1097 | 17.1 |
| Inceptisols | 12.86 | 9.83 | 1266 | 19.8 |

|  |  |  |  | Table 1: cont.... |
|---|---|---|---|---|
| Entisols | 21.14 | 16.16 | 1027 | 16.0 |
| Shifting sand | 5.32 | 4.07 | 82 | 1.3 |
| Rocky land | 13.02 | 10.00 | 176 | 2.7 |
| Glaciers, water bodies | 10.01 | 7.65 | 6 | 0.1 |

In addition, historically, communities first established on the alluvial plains and undulating lands that required low traction for management. With technological advances, this changed as evidenced by the increasing use of vertisols and aridisols. In the temperate parts of the world, alfisols and mollisols have high concentrations of people. The mollisols occupy about 6.9 % of the land surface and have about 6.7 % of the population on it. They are some of the best soils of the world but are mostly confined to the temperate countries. In the tropics, much of the population is associated with river terraces (entisols and inceptisols) and on ultisols. The ultisols and oxisols are problematic soils for low-input agriculture but, as demonstrated by the Brazilians, can be made productive with appropriate technology [8].

The gelisols of the Boreal zone have the lowest population density with about 2 persons per $km^2$ while the andisols (developed on volcanic pyroclastic materials) have the highest with more than 106 persons per $km^2$. Rwanda, Burundi, and Ituri province of eastern Zaire have the highest population densities in the world and this is followed by the volcanic areas of Southeast and East Asia. The ultisols and vertisols that dominate in the tropics have a population density of about 90 and 98 persons, respectively, while the mollisols and alfisols, the major grain producing regions of the temperate regions, have a density of 90 and 41 persons per $km^2$, respectively. Fragile systems such as those with histosols and aridisols have 18 and 20 persons per $km^2$, respectively, and though these are low, they are already threatening the sustainability of these systems. The largest extent of the histosols (organic soils) is in Canada. In the tropics, it is in Indonesia where shifting cultivation and very low input agriculture is destroying the ecosystem. The recent forest fires in Indonesia are partly due to this mismanagement. Historical land use studies have shown that populations have always sought the better soils for agriculture and, hence, the development of their communities. In more recent times, with advances in technology, particularly irrigation techniques, agriculture has moved into more fragile ecosystems. In the developing countries of the world, a burgeoning population has forced the poor landless also to move into fragile ecosystems or degrade the better resources of their countries. Land quality is a measure of the land to perform specific functions [9] and the features of each class are given in Table **2**.

**Table 2:** Properties of land quality classes.

| Land Quality Class | Properties |
|---|---|
| I | This is prime land. Soils are highly productive, with few management-related constraints. Soil temperature and moisture conditions are ideal for annual crops. Soil management consists largely of sensible conservation practices to minimize erosion, appropriate fertilization, and use of best available plant materials. Risk of sustainable grain crop production is generally < 20 %. |
| II& III | The soils are good and have few problems for sustainable production. However, and particularly for class II soils, care must be taken to reduce degradation. The lower resilience characteristics of class II soils make them more risky, particular for low-input grain crop production. However, their productivity is generally very high and, consequently, response to management is high. Conservation tillage is essential, buffer strips are generally required and fertilizer use must be carefully managed. Due to the relatively good terrain conditions, the land is suitable for national parks and biodiversity zones. Risk for sustainable grain crop production is generally 20-40 % but risks can be reduced with good conservation practices. |
| IV, V & VI | If there is a choice, these soils must not be used for grain crop production, particularly soils belonging to class IV. All 3 classes require important inputs of conservation management. In fact, no grain crop production must be contemplated in the absence of a good conservation plan. Lack of plant nutrients is a major constraint and so a good fertilizer use plan must be adopted. Soil degradation must be continuously monitored. Productivity is not high and so low-input farmers must receive considerable support to manage these soils or be discouraged from using them. Land can be set aside for national parks or as biodiversity zones. In the semi-arid areas, they can be managed for range. Risk for sustainable grain crop production is 40-60 %. |
| VII | These soils may only be used for grain crop production if there is a real pressure on land. They are definitely not suitable for low-input grain crop production; their low resilience makes them easily prone to degradation. They should be retained under natural forests or range and some localized areas can be used for recreational purposes. As in class V and VI, biodiversity management is crucial in these areas. Risk for sustainable grain crop production is 60-80% |

**Table 2: cont....**

| VIII& IX | These are soils belonging to very fragile ecosystems or are very uneconomical to use for grain crop production. They should be retained under their natural state. Some areas may be used for recreational purposes but under very controlled conditions. In class VIII, which is largely confined to the tundra and boreal areas, timber harvesting must be done very carefully with considerable attention to ecosystem damage. Class IX is mainly the deserts where biomass production is very low. Risk for sustainable grain crop production is >80 %. |
|---|---|

Land quality is assessed by 6 major functions [10, 11] each of which is equally important for human well-being:

1.  production of biomass through agriculture and forestry,

2.  protect the ground water and the food chain against pollution and maintaining biodiversity by filtering, buffering, and transformation activities,

3.  contribute to the preservation of the gene reserve by enabling the habitat for biota;

4.  provide the physical basis for infrastructural development, such as housing, industrial production, transport, dumping of refuse, sports, recreation, and others,

5.  serve as a source of raw materials, furnishing gravel, sand, clay, and other materials,

6.  preserve the geogenic and cultural heritage by concealing and protecting archaeological and paleontological materials.

All the functions are important for human well-being, but in the context of this study, the function that is most relevant is to sustain grain production and respond to cultural practices conducive to sustainable land management. Land quality is then assessed as the ability of soils to produce grain and the nine classes defined by Beinroth *et al.,* [9]. The global distribution of the nine land quality classes is depicted in the map of Fig. **1** and their respective areas are presented in Table **2**. Class I lands have ideal soils occurring in ideal climates for crop production and are characterized by high productivity, high response to management, and minimal limitations. They occupy only about 2.4 percent of the world's land surface but contribute more than 40 percent of global food and feed output. Over 90 percent of Class I soils are used for grain production, although in some countries (such as in Uruguay) they are used for grazing, perhaps because of the high labor costs associated with cultivation. Due to their productivity, most conservation investments are also found on this class of soils because of the assured rewards of sustainable land management.

The 9.5 percent of the global land resources in LQ classes II and III have minor limitations that generally are easily corrected and that do not pose permanent restrictions to the use of the land. Most of these lands are in the temperate regions of the world where the climate is moderate, with minimal extremes of rainfall or temperature. These soils respond well to management and the positive effects of appropriate management persist for long periods. Unlike class I soils that are dominantly in the tropics, class II and III soils have a wider distribution.

Land quality classes IV, V, and VI together cover 34 percent of the world's land area, largely in the tropics and support about 54% of the population (Table **3**).

**Table 3:** Area (million km$^2$) in land quality class with estimate of population (million) in each class.

| Land Quality Class | Area | % | Nr People | % |
|---|---|---|---|---|
| I | 3.11 | 2.38 | 388 | 6.1 |
| II | 6.51 | 4.98 | 908 | 14.2 |
| III | 5.95 | 4.55 | 306 | 4.8 |
| IV | 5.17 | 3.95 | 753 | 11.8 |
| V | 21.60 | 16.51 | 1900 | 29.7 |
| VI | 17.42 | 13.32 | 777 | 12.1 |
| VII | 11.79 | 9.01 | 735 | 11.5 |
| VIII | 21.83 | 16.69 | 119 | 1.9 |
| IX | 35.19 | 28.59 | 719 | 11.2 |

These soils have a range of constraints, from high ambient temperatures that reduce germination rates to low nutrient availability that limits biomass production of annual crops (however, some of these lands are niches for specific land use such as plantations of rubber, oil palm, and cocoa in the tropics). Due to their extent in the tropics (Table **4**) where much of the world's population resides, these lands are of particular significance and require additional research and development initiatives. High population densities, coupled with the prevalence of low-input agriculture, make these lands highly vulnerable to human-induced degradation and desertification

**Table 4:** Percent land area in major biomas as a function of land quality.

| Biomas | Land Quality Class (percent of ice-free land surface) | | | | | | | | |
|---|---|---|---|---|---|---|---|---|---|
| | I | II | III | IV | V | VI | VII | VIII | IX |
| Tundra | | | | | | | | 15.62 | |
| Boreal | | | 2.03 | 0.67 | 0.50 | 3.05 | 2.63 | 1.08 | 0.07 |
| Temperate | 2.14 | 2.55 | 0.70 | 1.31 | 4.76 | 1.66 | 2.01 | | 0.15 |
| Mediterranean | | | 0.30 | 0.15 | 1.35 | 0.08 | 0.65 | | 0.03 |
| Desert | | | | | | | 1.42 | | 28.19 |
| Tropical | 0.25 | 2.43 | 1.51 | 1.83 | 9.90 | 8.53 | 2.31 | | 0.16 |
| Total | 2.38 | 4.98 | 4.55 | 3.95 | 16.51 | 13.32 | 9.01 | 16.59 | 28.59 |

Land quality class VII soils occupy about 9 percent of global land area and comprise shallow soils, those with high salt concentrations and those with high organic matter. The first are generally excluded in most assessments of suitable land for agriculture. The peat lands are included in this group due to their fragility and hence the inherent dangers associated with their use. The peat lands may be permanently lost via drainage, as has happened in many parts of Southeast Asia. Their uniqueness stems from the fact that they perform specific roles as wetlands and they are also the most efficient sequesters of organic carbon. These lands support about 11.5 % of the population and this amount is constantly increasing in the developing parts of the world due to incursions by the landless.

Land quality class VIII lands, covering 17 percent of the world's land surface, have low temperatures and/or occur on steep slopes, implying that they are generally unsuitable for agriculture, even though their net primary productivity (NPP) may be moderately high. Included in this class are the extremely fragile peat lands of the high latitudes. Perturbation of this ecosystem through land clearing or climate change results in destruction of the permafrost with accompanying oxidation of the peat. Land quality class IX occupies about 29 percent of the world's land surface. This group, comprising soils with inadequate moisture to support most annual crops (and also rocky land and sand dunes), has a very low net primary productivity. Nevertheless, this class includes deep soils that, given high solar radiation in summer, are highly productive under irrigation. Efficient use of water is crucial to the management of such soils as their resilience, for example when degraded by salinization, is generally very low classes VII, VIII, and IX lands, because of their fragility and very high risks both for ecosystem integrity and sustainable agriculture, should be free of human intervention. Large areas of the Taiga and tropical peat forests are currently threatened by shifting cultivation, but their highest value may be in the provision of environmental services, such as biodiversity, carbon sequestration, and water quality enhancement.

## SUSTAINING FOOD PRODUCTION

During the next 2 decades, trends in population, income, and urbanization are projected to raise world demand for cereals and tubers by 40% and for meat by about 60% [12]. The ability to become sustainable varies, depending on the natural resources available and their conditions in each country. As the increased production to meet this demand will have to come from increased productivity, at least in Asia, the state of the resource base determines the ability of each country to meet its food and fiber needs.

There are a number of estimates on global population supporting capacity, and Eswaran *et al.,* [4] employed     the concept of land quality and relative grain producing capacity to make such estimates. The assessment is used here only to show magnitudes and geographic areas of concern. By merely using class I land it is possible to support the current world population in an idealistic society, where everything is shared and there is no problem for access to food. The same ideal global society can support more than 30 billion persons if it uses all the land from class I through V. This is the absolute maximum. In a more pragmatic world, class I and II lands together can support about

10 billion people. The conclusion of this study, similar to the assessment by Greenland *et al.,* [13] is that famine and starvation of people of some countries is not because of the inability of global land resources to produce the necessary food. Class I and II lands do not occur in all countries. Some countries, such as Afghanistan and Pakistan, have an insignificant amount of class I through IV lands, and hence have to rely on risky irrigation of class VII lands to meet their food needs. Kuwait and Japan, with similar land resource problems, rely on other pathways to provide the food. A more rigorous analysis was made by Beinroth *et al.,* [9]    for Asia and they showed that, for Asia as a whole, food insecurity is an acute problem. In fact, for some countries the Malthusian prophecy is rapidly becoming a reality. The study also revealed the relative scarcity of prime agricultural land in Asia and the resulting imperative to preserve these areas for food production and optimize the land use of the remaining areas.

It is evident that not all countries are endowed with good-quality land resources to produce all the food they need. The situation is aggravated by the fact that in countries characterized by a predominance of limited resource, farmers who are caught in the poverty spiral described by McCowan and Jones [14] land degradation and desertification are further reducing the capacity of soils to produce.

Although soils play an important role in sustaining food production, the challenge is to enable this in the socioeconomic and political context. It must be recognized that each country has a range of soils that vary in productivity and fragility. Optimizing land use in countries with a dominance of agrarian population is difficult. An important policy challenge for both industrialized and developing countries is to find ways to maintain and enhance food production, while seeking both to improve the positive functions and to eliminate the negative ones, so improving the overall sustainability of rural livelihoods and economies. From 1948, the total land area under cultivation shows a gradual decline in United States. The inputs for production have declined, but with an increase in the outputs. Productivity (a ratio of outputs to inputs) shows a systematic increase with time. The efficiency of production (productivity per unit land) is more revealing. It shows a linear increase until about 1982 when grain crops had a high price. With the high price, less-suitable land was brought under production with a concomitant decline in efficiency of production. However, about 1982, the Conservation Reserve Program, whereby farmers were rewarded for setting aside unproductive land, came into effect. Since then efficiency of production has increased almost geometrically. This is an illustration of the impact of enabling conservation policies. Similar examples are also available in Europe, but such policies do not exist in many developing countries as a consequence of which land degradation is rampant.

## CONCLUSION

The ability of the land to feed and clothe people and to maintain ecological functions is being impeded by demographics. In addition to these population-linked issues are others, which are human-induced and represent a new generation of global environmental problems. The global land area that is generally free of constraints for most agricultural uses is unequally spread around the globe with a larger portion in the temperate countries of the world. In addition to poorer land quality in tropical regions, land degradation is also well entrenched, aggravating food security. There are 11.9 million $km^2$ of such lands and about 1.4 billion people are involved and most of these areas are in the developing countries.

Food security then becomes a major issue in those countries that are not blessed with good land resources or those who have degraded or are degrading their resources. Countries of the developing parts of the world have to make a conscious decision to better manage their land resources. The paradigm shift that poorer countries need to make to sustain food production is to implement holistic and sustainable land management programs by adopting technologies that have already been validated in other parts of the world. To assure sustained use of soil resources:

- Research investments must contribute to new knowledge and more productive means of food production,

- An active program of assessment and monitoring of land degradation must be institute to provide accurate and unbiased information,

- A proactive commitment to sustainability must be made, partly through wise land-use planning and implementation, to ensure that biodiversity is maintained and environments are preserved and protected,

- Appropriate national and international policy environments must exist to enable access to food through a fair and equitable market system so that countries can capitalize on niches,

- It must be recognized that the human carrying capacity of the land is not merely a national problem, but a global one, since it impacts every aspect of human society and is strongly linked to the soil resources.

Finally, the environmental and human health effects of mismanagement of land are wide-ranging and include: (i) sealing of land, (ii) chemical pollution contaminating water and harming wildlife and human health; (iii) excessive use of fertilizers such as nitrate and phosphate fertilizers, livestock wastes, and silage effluents contaminating water, and thereby contributing to algal blooms, deoxygenation, fish deaths, and hazards for recreation; (iv) soil erosion disrupting water courses, and runoff from eroded land causing flooding and damage to housing and natural resources and resulting in billions of dollars of damage; (v) harm to the food-chain exposed to toxic residues and microorganisms in foods; and (vi) contamination of the atmospheric environment by methane, nitrogen oxide, and ammonia derived from livestock, their manure, and fertilizers. The social cost of these is high, but more important is the loss of natural capital that cannot be replenished.

Properly managed, land also delivers valued nonfood functions, many of which cannot be produced by other economic sectors. The aesthetic value, recreation and amenity, water accumulation and supply, nutrient recycling and fixation including carbon sequestration, wildlife, including agriculturally beneficial organisms, and storm protection and flood control are examples. Positive social externalities include provision of jobs, contribution to the local economy, and to the social fabric of rural communities.

To sum up, soils are crucial to sustain food production; and to enable soils to perform their functions, efforts (based on solar energy, plant activity, and soil restoration) must be made to protect and conserve the soil resource.

## REFERENCES

[1]　Virmani SM, Katyal JC. Eswaran H, Abrol I. Stressed Agroecosystems and Sustainable Agriculture. New Delhi: Oxford and IBH, 1994.

[2]　Cleaver KM, Schreiber GA. Reversing the Spiral: The Population, Agriculture, and Environment Nexus in Sub-Saharan Africa. Washington, DC: World Bank, 1994.

[3]　Eswaran H. Soil taxonomy and agrotechnology transfer. In: van Cleemput O, Ed. Soils for Development. ITC Ghent, Belgium. Publication Series 1989; 1: 9-28.

[4]　Eswaran H, Beinroth F, Reich P. Global land resources and population supporting capacity. Am. J. Alternative Agric 1999; 14: 129-36.

[5]　Durning AB. Poverty, and the environment: Reversing the downward spiral. Worldwatch Institute, Washington DC. 1989.

[6]　Soil Survey Staff. Soil Taxonomy: A Basic System of Soil Classification for making and Interpreting Soil Surveys. 2nd edition. U.S. Dept. Agric. Handbook 436. Government Printing Office, Washington, DC. 1999.

[7]　Tobler W, Deichmann U, Gottsegen J, Maloy K. NCGIA Technical Report TR95-6 "The global demography project". Nat. Center for Geographic Information and Analysis. Univ. of California, CA. 1995.

[8]　Buol SW, Eswaran H. Assessment and conquest of poor soils. In: Maranville JW et al., Eds. Adaptation of plants to soil stresses. Intsormil Publication, 1994; pp. 17-27.

[9]　Beinroth F, Eswaran H, Reich P. Land quality and food security in Asia. 2nd International Conference on Land Degradation. Khon Kean. Thailand. CD ROM, Department of Land Development, Bangkok, Thailand. January 2001.

[10]　Blum WEH. Agriculture in a sustainable environment-a holistic approach. Int. Agrophysics 1998; 12: 13-24.

[11]　Blum WEH. The role of soils in sustaining society and the environment: realities and challenges for the 21st century. 17th World Congress of Soil Science, Bangkok/Thailand. Keynote lectures, 2002; pp. 66-86.

[12]　Pinstrup-Andersen P, Pandya-Lorch R, Rosegrant MW. World Food Prospects: Critical Issues for the Early Twenty-First Century. Intl. Food Policy Research Inst. Washington, DC. 1999.

[13]　Greenland DJ, Gregory Pl, Nye PH. Land Resources: On the Edge of the Malthusian Precipice? CAB International, New York. 1998; pp.180.

[14]  McCowan RL, Jones RK. Agriculture of semi-arid eastern Kenya: problems and possibilities. In: Probert M, Ed. A Search for a Strategy for Sustainable Dryland Cropping in Semi-arid Eastern Kenya. AClAR, Canberra, Australia. 1992; pp. 8-15.

# GLOSSARY

## 2,4-D

2,4-Dichlorophenoxyacetic acid (2,4-D) is a common systemic herbicide used in the control of broadleaf weeds.One of the representative of the class of Hormonal herbicide

## Abscisic acid (ABA)

Abscisic acid (ABA), also known as abscisin II and dormin, is a plant hormone with inhibiting functions. ABA functions in many plant developmental processes, including bud dormancy; it is degraded by the enzyme, (+)-abscisic acid 8'-hydroxylase.

## Absorption

Absorption, in chemistry, is a physical or chemical phenomenon or a process in which atoms, molecules, or ions enter some bulk phase - gas, liquid or solid material.

## Actinomycetes

Actinomycetes are a diverse and a large group of gram positive filamentous and/ or branching bacilli.

## Adaptation

Adaptation is the process whereby a population becomes better suited to its habitat. This process takes place over many generations, and is one of the basic phenomena of biology

## Adventitous root

Adventitious roots arise out-of-sequence from the more usual root formation of branches of a primary root, and instead originate from the stem, branches, leaves, or old woody roots.

## Agricultural land

Agricultural land (also agricultural area) denotes the land suitable for agricultural production, both crops and livestock.

## *Agrobacterium tumefaciens*

*Agrobacterium tumefaciens* is the causal agent of crown gall disease (the formation of tumors) in over 140 species of dicot. Plasmides from *A. tumefaciens* is used sometimes as vectors in plant gene engineering.

## Agronomic crops

Agronomic crops typically involve a crop that is grown for grain, feed, or for processing into oil, starch, protein and flour.

## Agrosol

Artificial composition of soil elements used for agricultural purposes.

## Albino

Person or animal lacking normal pigmentation, with the result being that the skin and hair are abnormally white or milky and the eyes have a pink or blue iris and a deep-red pupil. A plant that lacks chlorophyll.

## Alkaloid

Alkaloids are naturally occurring chemical compounds containing basic nitrogen atoms. The name derives from the word alkaline and was used to describe any nitrogen-containing base. They are secondary substances in the plant metabolism

## Alfisol

Alfisols are a soil order in USDA soil taxonomy. Alfisols form in semiarid to humid areas, typically under a hardwood forest cover.

## Allelopathogenes

Secondary substances produced by some plants which have a strong influence on the other plant species

## Allelopathy

Allelopathy is a biological phenomenon that is characteristic of some plants, algae, bacteria, coral and fungi by which they produce certain biochemical's that influence the growth and development of other organisms.

## Amino acids

Amino acids are primary products of metabolism. Its molecules containing an amine group, a carboxylic acid group and a side chain that varies between different amino acids.

## Ammonification

When a plant dies, an animal dies, or an animal expels waste, the initial form of nitrogen is organic. Bacteria, or in some cases, fungi, convert the organic nitrogen within the remains back into ammonium ($NH_4^+$), a process called ammonification or mineralization.

## Ammonium

The ammonium cation (also known as ionized ammonia due to its electrical charge) is a positively charged polyatomic cation of the chemical formula $NH^+_4$.

## Anabolism

Anabolism is the set of metabolic pathways that construct molecules from smaller units

## Anthocyanin

Anthocyanins are water-soluble vacuolar pigments in plants that may appear red, purple, or blue according to pH.

## Anthranilic acid

Anthranilic acid is the organic compound with the formula $C_6H_4(NH_2)COOH$. One of the primary products of amino acid- triptophane metabolism. Anthranilic acid is used as an intermediate for production of dyes, pigments, and saccharin

## Atomic absoption spectrophometer

In analytical chemistry, atomic absorption spectroscopy is a technique for determining the concentration of a particular metal element in a sample

## Atomic emission spectrophometer

Atomic emission spectroscopy (AES) is a method of chemical analysis that uses the intensity of light emitted from a flame, plasma, arc, or spark at a particular wavelength to determine the quantity of an element in a sample.

## Auxin

Auxins are a class of plant growth substance and morphogens (often called phytohormone or plant hormone).

**Auxin oxidase**

peroxidase and co- factors which converts indolic auxin ( indolyl-acetic acid ) to indolyl carbonic acid

**Benzylaminopurine (BA)**

Benzylaminopurine, benzyl adenine or BAP is a first-generation synthetic cytokinin which elicits plant growth and development responses, setting blossoms and stimulating fruit richness by stimulating cell division.

**Bacillus**

*Bacillus* is a genus of rod-shaped bacteria and a member of the division Firmicutes. *Bacillus* species are obligate aerobes, and test positive for the enzyme catalase, ubiquitous in nature.

**Bacteria**

The bacteria are a large group of unicellular microorganisms. Typically a few micrometers in length, bacteria have a wide range of shapes, ranging from spheres to rods and spirals.

**Bacteria cenosis**

Complex of bacteria species on a certain space of plant or soil

**Bark**

Bark is the covering of the stems of woody plants, like trees, protects the tree. Bark of different plants and trees can look very different; it can be rough or smooth and can have different colors. Bark contains conducting vessels for the transport of photosynthetic products

**Bioflora**

Complex of biological organisms like plants, fungi and bacteria.

**Biological activity**

Pharmacological or biological activity is an expression describing the beneficial or adverse effects of a drug on living matter of living organisms

**Biological rhythm**

Periodic biological fluctuation in an organism corresponding to and in response to periodic environmental change, such as day and night or high and low tide.

**Biological test**

Determining of the biological function of a substance by examining its influence on the growth of organisms

**Biomass**

Biomass, a renewable energy source, is biological material derived from living, or recently living organisms, such as wood, waste, and alcohol fuels.

**Biomass accumulation**

The deposition of proteins, fats, carbohydrates and some secondary substances in the plant organs which forms the body of the plat

**Biomass production**

The biosynthetic processes of photosynthesis, transport of assimilate and deposit it in the plant organs

**Bio-minesoil**

Fabricated soil which covers mining substrates

**Biosphere**

The biosphere is the global sum of all ecosystems. It can also be called the zone of life on Earth.

**Biosynthesis**

Biosynthesis (also called biogenesis) is an enzyme-catalyzed process in cells of living organisms by which substrates are converted to more complex products.

**Biotechnology**

Biotechnology is technology based on biology, agriculture, food science, and medicine. Modern use of the term usually refers to genetic engineering as well as cell- and tissue culture technologies.

**Biotest**

A method for assessing the effect of a compound, technique, or procedure on an organism. Synonym: biological assay.

**Boron**

Boron is the chemical element with atomic number 5 and the chemical symbol B. B. is a micro-element in plant nutrition

**Branch root**

Another name for lateral root. Originates from the pericycle.

**Bud**

In botany, a bud is an undeveloped or embryonic shoot and normally occurs in the axil of a leaf or at the tip of the stem.

**C-3 plants**

Plants that survive solely on $C_3$ fixation ($C_3$ plants) tend to thrive in areas where sunlight intensity is moderate, temperatures are moderate, carbon dioxide concentrations are around 200 ppm or higher, and ground water is plentiful.

**C-4 plants**

$C_4$ carbon fixation is one of three biochemical mechanisms, along with $C_3$ and CAM photosynthesis, functioning in land plants to "fix" carbon dioxide for sugar production through photosynthesis.

**Caffeic acid**

Caffeic acid, which is unrelated to caffeine, is biosynthesized by hydroxylation of coumaroyl ester of quinic ester.

**Calcium**

Calcium is the chemical element with the symbol Ca and atomic number 20. Macro- element in the plant nutrition

**Callus**

Callus is a group of cells which produce tumor –like tissues.A callus cell culture is usually sustained on gel media, much in the same manner as bacteria are grown. Sufficient media consists of agar and the usual mix of macronutrients and micronutrients for the given cell type.

**Carbohydrate**

A carbohydrate is an organic compound with general formula $C_m(H_2O)_n$, that is, consisting only of carbon, hydrogen and oxygen, the last two in the 2:1 atom ratio.

**Carbon**

Carbon is the chemical element with symbol C and atomic number 6.

**Carbon dioxide**

Carbon dioxide (chemical formula $CO_2$) is a chemical compound composed of two oxygen atoms covalently bonded to a single carbon atom.

**Carbon reduction**

The process of the photosynthetic transformation of carbon dioxide to primary organic products.

**Cambium**

In botany this is a layer or layers of tissue, also known as lateral meristems, that are the source of cells for secondary growth.

**Cambial zone**

Cambium containing area.

**Carotenoid**

Carotenoids are organic, lipid soluble pigments naturally occurring in plants and some other photosynthetic organisms like algae , some types of fungus and some bacteria .

**Cell**

The cell is the basic structural and functional unit of all known living organisms.

**Cellulase**

Cellulase refers to a class of enzymes produced chiefly by fungi, bacteria, and protozoans that catalyze the cellulolysis (or hydrolysis) of cellulose.

**Cellulose**

Cellulose is an organic compound with the formula $(C_6H_{10}O_5)_n$, a polysaccharide consisting of a linear chain of several hundred to over ten thousand $\beta(1{\rightarrow}4)$ linked D-glucose units.

**Chemical Signaling**

Hormonal substances which regulate metabolic streams and organogenesis.

**Chemical signals**

Hormonal factors add some mineral ions which participate in the processes of morphogenesis and growth

### Chemo-taxonomy

Chemotaxonomy (from chemistry and taxonomy), also called chemosystematics, is the attempt to classify and identify organisms (originally plants), according to demonstrable differences and similarities in their biochemical compositions.

### Chloride

The chloride ion is formed when the element chlorine picks up one electron to form an anion (negatively-charged ion) $Cl^-$.

### Chlorophyll

Chlorophyll is a green pigment found in most plants, algae, and cyanobacteria.

### Chloroplast

Chloroplasts are organelles found in plant cells and other eukaryotic organisms that conduct photosynthesis.

### Chromosome

A chromosome is an organized structure of DNA and protein that is found in cells. It is a single piece of coiled DNA containing many genes, regulatory elements and other nucleotide sequences.

### Cinnamic acid

Cinnamic acid has the formula $C_6H_5CHCHCOOH$ and is a white crystalline acid, which is slightly soluble in water.

### Clone

Cloning in biology is the process of producing populations of genetically-identical individuals that occurs in nature when organisms such as bacteria, insects or plants reproduce asexually.

### Clover

Clover (*Trifolium*), or trefoil, is a genus of about 300 species of plants in the pea family Fabaceae.

### Compositing toilet leachate

Liquid faze of composting products rich of nitrogen products.

### Copper

Chemical element with the symbol Cu (Latin: *cuprum*) and atomic number 29.

### Coumarin

Coumarin is a secondary chemical compound found in many plants, notably in high concentration in the tonka bean (*Dipteryx odorata*), vanilla grass (*Anthoxanthum odoratum*), woodruff (*Galium odoratum*), mullein (*Verbascum* spp.), and sweet grass (*Hierochloe odorata*).

### Crop

A crop is the annual or season's yield of any plant that is grown in significant quantities to be harvested as food, as livestock fodder, fuel, or for any other economic purpose.

### Coumaric acid,

Coumaric acids are organic compounds that are hydroxy derivatives of cinnamic acid.

**Cutting**

Cutting is the separation of a physical object, or a portion of a physical object, into two portions, through the application of an acutely directed force.

**Cytokinins**

Cytokinins (CK) are a class of nitrogen containing plant growth substances (plant hormones) that promote cell division and leaves greening.

**Decidious**

Deciduous means falling off at maturity or tending to fall off and is typically used in reference to trees or shrubs that lose their leaves seasonally and to the shedding of other plant structures such as petals after flowering or fruit when ripe.

**Decomposer**

Decomposers (or saprotrophs) are organisms that eat the dead or decaying organisms, and in doing so carry out the natural process of decomposition.

**Denitrification**

Denitrification is a microbially facilitated process of dissimilatory nitrate reduction that may ultimately produce molecular nitrogen ($N_2$) through a series of intermediate gaseous nitrogen oxide products.

**Desert**

A desert is a landscape or region that receives almost no precipitation.

**Desertification**

Desertification is the extreme deterioration of land in arid and dry sub-humid areas due to loss of vegetation and soil moisture; desertification results chiefly from man-made activities and influenced by climatic variations.

**Dicots**

Dicotyledons, or "dicots", are a name for a group of flowering plants whose seed typically has two embryonic leaves or cotyledons.

**DNA**

Deoxyribonucleic acid (DNA) is a nucleic acid that contains the genetic instructions used in the development and functioning of all known living organisms and some viruses.

**Dormancy**

Dormancy is a period in an organism's life cycle when growth, development, and (in animals) physical activity is temporarily stopped. This minimizes metabolic activity and therefore helps an organism to conserve energy

**Ecological engineering**

Ecological engineering is an emerging of study integrating ecology and engineering, concerned with the design, monitoring and construction of ecosystems.

**Ecological indicators**

Ecological indicators are used to communicate information about ecosystems and the impact human activity has on ecosystems to groups such as the public or government policy makers.

## Ecological modeling

Ecosystem models, or ecological models, are mathematical representations of ecosystems.

## Ecology

Ecology is the interdisciplinary scientific study of the interactions between organisms and the interactions of these organisms with their environment.

## Ecosystem

The term ecosystem refers to the combined physical and biological components of an environment.

## Electromagnetic spectrum,

The electromagnetic spectrum is the range of all possible frequencies of electromagnetic radiation

## Ethylene

Ethylene is the chemical compound with the formula $C_2H_4$. It It is extremely important in industry and also has a role in biology as a hormone of fruit ripening.

## Etiolation

Etiolation occur when plants are grown in either partial or complete absence of light, and is characterized by long, weak stems; smaller, sparser leaves due to longer internodes; and a pale yellow color (chlorosis).

## Evolution

In biology, evolution is change in the genetic material of a population of organisms through successive generations.

## Expansion

Increase in size of the cells or tissues.

## Fabricated soil (FS)

Artificialy composed soil analogues with high and long lasting fertility

## Fabricated soil experiment

The manipulation with the components of fabricated soil for the increase of its fertility

## Fermentation

Fermentation is the process of deriving energy from the oxidation of organic compounds, such as carbohydrates, and using an endogenous electron acceptor, which is usually an organic compound.

## Fibers

Fiber, also spelled fibre, is a class of materials that are continuous filaments or are in discrete elongated pieces, similar to lengths of thread. They are very important in the biology of both plants and animals, for holding tissues together.

## Flavanoid biosynthesis

Flavonoids are synthesized by the phenylpropanoid metabolic pathway in which the amino acid phenylalanine is used to produce 4-coumaroyl-CoA. Often flavanoid have different type of pigmentations

## Flavanol

Flavan-3-ols (sometimes referred to as flavanols) are a class of flavonoids that use the 2-phenyl-3,4-dihydro-2H-chromen-3-ol skeleton.

## Fungus

A fungus is a member of a large group of eukaryotic organisms that includes microorganisms such as yeasts and molds, as well as the more familiar mushrooms. Fungus is mostly heterotrophs.

## Gene

A gene is the basic unit of heredity in a living organism.

## Genome

In modern molecular biology, the genome is the entirety of an organism's hereditary information. It is encoded either in DNA or, for many types of virus, in RNA.

## Genotype

The genotype is the genetic constitution of a cell, an organism, or an individual (i.e. the specific allele makeup of the individual) usually with reference to a specific character under consideration.

## Germination

Germination is the process in which a seed or spore emerges from a period of dormancy.

## Gibberellins

Gibberellins (GAs) are plant hormones that regulate growth and influence various developmental processes, including stem elongation, germination, dormancy, flowering, sex expression, enzyme induction, and leaf and fruit senescence.

## Global soil resources

The part of fertile lands, often in the agricultural use on the continents

## Grassy plant

Annual and perennial herbs.

## Green algae

The green algae (singular: green alga) are the large group of algae from which the embryophytes (higher plants) emerged

## Grey water system

The system of cleaning of the sink water for its re - use

## Growth

Normal process of increase in size of an organism as a result of accretion of tissue similar to that originally present.

## Growth ring

A layer of wood formed in a plant during a single period of growth.

## Growth and development

Growth is defined as an increase in size; development is defined as a progression toward maturity.

## Growth regulation

System of hormones and its co-factors for the differentiation of tissues and organs of organism.

## Growth regulator

Plant hormones (also known as phytohormones) are chemicals that regulate plant growth, which, are termed also 'plant growth substances'

## Hard wood

The term 'hardwood' is used to describe wood from angiosperm trees (more strictly speaking non-monocot angiosperm trees).

## Hemicellulose

A hemicellulose can be any of several heteropolymers (matrix polysaccharides) present in almost all plant cell walls along with cellulose.

## Herb

An herb is a plant that is valued for flavor, scent, or other qualities

## Herbicide

An herbicide is a substance used to kill unwanted plants.

## Hormone

A hormone is a chemical released by one or more cells that affects cells in other parts of the organism.

## Housing

The process of the shelter creation and maintenance

## Humus

Humus is degraded organic material in soil, which causes some soil layers to be dark brown or black

## Hyroxycinnamic acid

Hydroxycinnamic acids are a class of polyphenolic compounds that are hydroxy derivatives of cinnamic acid.

## Indolacetonitrile

One of the precursors of indolic auxin - indolyl- acetic acid (IAA).

## Indolyl-3-acetic acid (IAA)

Indole-3-acetic acid, also known as IAA, is a heterocyclic compound that is an phytohormone called auxins.

## Inhibitor

Something that restrains, blocks, or suppresses.

## Ion chromatography

Ion-exchange chromatography (or ion chromatography) is a process that allows the separation of ions and polar molecules based on their charge.

## Iron

Iron is a metallic chemical element with the symbol Fe and atomic number 26.

## Krebs cycle

The citric acid cycle, also known as the tricarboxylic acid cycle (TCA cycle), the Krebs cycle, is a series of enzyme catalyzed chemical reactions, which is of central importance in all living cells that use oxygen as part of cellular respiration.

## Land biodiversity

Various living organisms as elements of eco-system.

## Leaf inhibitory properties

Some leaves which are enriched by secondary substances and abscisic acid which block the processes of stem growth and buds morphogenesis

## Legume

A legume in botanical writing is a plant in the family Fabaceae (or Leguminosae), or a fruit of these specific plants

## Lettuce

Lettuce (*Lactuca sativa*) is a temperate annual or biennial plant of the daisy family Asteraceae.

## Lignin

Lignin is a complex chemical compound most commonly derived from wood, and an integral part of the secondary cell walls of plants and some algae. Chemicaly lignin is a carbon polymer. Lignin units – monomers are oxidised phenolics

## Lignification

The process of lignin biosynthesis and deposition during the cells and tissues differentiation.

## Macroelements

Minerals like nitrogen, phosphorus, potassium, sulphur, magnesium, iron amd oth required for optimal functioning of the body; dietary requirements for minerals range from molar to trace amounts/day; some– eg, nickel, tin, and vanadium, may be required by some plants or animals, but are not known to have a role in humans

## Magnesium

Magnesium is a chemical element with the symbol Mg, atomic number 12 and common oxidation number +2.

## Manganese

Manganese is a chemical element, designated by the symbol Mn. It has the atomic number 25. In plant bodies is functioning as micro- element

### Mesophyll

The soft chlorophyll-containing tissue of a leaf between the upper and lower layers of epidermis: involved in photosynthesis

### Mevalonic acid

Mevalonic acid is a precursor in the biosynthetic pathway, known as the mevalonate pathway which produces terpenes and steroids.

### Microbial activity

Metabolic activity of some micro-organisms in different regions of eco- systems.

### Microbial component

Complex of microorganisms in the eco-system

### Microelements

Trace elements required for plant growth and development like manganese, boron, zink and other.

### Micronutrients

Micronutrients are nutrients needed throughout life in small quantities.

### Molybdenum

Molybdenum is a chemical element with the symbol Mo and atomic number 42.

### Monocots

Monocotyledons or monocots are one of two major groups of flowering plants (angiosperms) that are traditionally recognized, the other being dicotyledons or dicots.

### MS medium

Murashige and Skoog medium or (MS medium) is a plant growth medium used in the laboratories for cultivation of plant cell culture.

### Mutation

Mutations are changes in the DNA sequence of a cell's genome and are caused by radiation, viruses, transposons and mutagenic chemicals, as well as errors that occur during meiosis or DNA replication.[

### Mycelia

Mycelium (plural mycelia) is the vegetative part of a fungus, consisting of a mass of branching, thread-like hyphae.

### Naphthaleneacetic acid (NAA)

1-naphthaleneacetic acid - synthetic auxin (organic compound and plant hormone)

### Naringenin

Naringenin is a flavonoid that is considered to have a bioactive effect on human health as antioxidant, free radical scavenger, anti-inflammatory, carbohydrate metabolism promoter, and immune system modulator.

### Natural phenolic inhibitors

Some phenolics like coumarins or coumaric acids which inhibit growth processes

**Nitrate ion**

In inorganic chemistry, a nitrate is a salt of nitric acid with an ion composed of one nitrogen and three oxygen atoms ($NO^{-3}$).

**Nitrification**

Nitrification is the biological oxidation of ammonia with oxygen into nitrite followed by the oxidation of these nitrites into nitrates.

**Nitrifying bacteria**

Nitrifying bacteria are chemoautotrophic bacteria that grow by consuming inorganic nitrogen compounds.[

**Nitrogen**

Nitrogen is a chemical element that has the symbol N, the atomic number of 7.

**Nitrogen fixation**

Nitrogen fixation usually refers to the biological process by which nitrogen ($N_2$) in the atmosphere is converted into ammonia.

**Nitrogen fixing organisms**

Soil nutrition formulas and beneficial soil-organisms such as nitrogen-fixing bacteria and mycorrhiza.

**Non-agricultural land**

Element of eco- system free from agricultural use

**Nutrient cycle**

Transfer of nutrients from one part of an ecosystem to another. Trees, for example, take up nutrients such as calcium and potassium from the soil through their root systems and store them in leaves.

**Ontogeny**

Ontogeny describes the origin and the development of an organism from the fertilized egg to its mature form.

**Organs**

A collection of tissues joined in structural unit to serve a common function.

**Osmosis**

Osmosis is the diffusion of water through a semi-permeable membrane. More specifically, it is the movement of water across a semi-permeable membrane from an area of high water potential (low solute concentration) to an area of low water potential (high solute concentration).

**Oxidases**

An oxidase is any enzyme that catalyzes an oxidation/reduction reaction involving molecular oxygen ($O_2$) as the electron acceptor. In these reactions, oxygen is reduced to water ($H_2O$) or hydrogen peroxide ($H_2O_2$).

**Oxidation**

In chemistry, the oxidation state is an indicator of the degree of oxidation of an atom in a chemical compound.

## Oxidative phosphorylation

Oxidative phosphorylation is a metabolic pathway that uses energy released by the oxidation of nutrients to produce adenosine triphosphate (ATP).

## PAL

phenyl alanine ammonium lyase- enzyme which transforms phenyl- alanine by desamination process to cinnamic acid, precursor of phenolics.

## Pericycle

The pericycle is a cylinder of parenchyma cells that lies just inside the endodermis and is the outer most part of the stele of plants.

## Periderm

In smaller stems and on typically non woody plants, sometimes secondary covering forms called the periderm, which is composed of cork (phellem), the cork cambium (phellogen), and the phelloderm.

## pH

pH is a measure of the acidity or basicity of a solution.

## Phloem

Vascular plants (also known as tracheophytes or higher plants) are those plants that have lignified tissues for conducting water, minerals, and photosynthetic products through the plant.

## Phloem ray

A vascular ray extending into or located entirely within the secondary phloem.

## Phenol

Phenol, also known as carbolic acid, is a toxic, white with a slightly pink tinge, crystalline solid. Its chemical formula is $C_6H_5OH$ and its structure is that of a hydroxyl group (-OH) bonded to a phenyl ring, making it an aromatic compound.

## Phenolic compounds

The phenols compounds include a large group of several hundred chemical compounds, known as polyphenolics, that affect the taste, color and mouthfeel of wine.

## Phenolic cycle

Transformation of phenolic substances in the living plant and its residue

## Phenolic inhibitors

Phenolic derivatives which inhibit seeds germination and plant growth

## Phenolic secretion

The ability of roots to secrete phenolic substances or the ability of phenolics to be evacuated from plant residues to water and soil

## Phenotype

A phenotype is any observable characteristic or trait of an organism: such as its morphology, development, biochemical or physiological properties, or behavior.

## Phosphate

A phosphate, an inorganic chemical, is a salt of phosphoric acid. In organic chemistry, a phosphate, or organophosphate, is an ester of phosphoric acid.

## Phosphorous

Phosphorus is the chemical element that has the symbol P and atomic number 15.

## Photomorphogenesis

In developmental biology, photomorphogenesis is light-mediated development. The photomorphogenesis of plants is often studied by using tightly-frequency-controlled light sources to grow the plants.

## Photosynthesis

Is a process that converts carbon dioxide into organic compounds, especially sugars, using the energy from sunlight.

## Phytochrome

Phytochrome is a photoreceptor, a pigment that plants use to detect light. It is sensitive to light in the red and far-red region of the visible spectrum.

## Phytohormones

Plant hormones (also known as phytohormones) are chemicals that regulate plant growth, which, in the UK, are termed 'plant growth substances'.

## Plant diversity

The variability among plants on the earth, including the variability within and between species and within and between ecosystems.

## Plant management

The way of creation and treatment of plant cenosis

## Plasma spectrophotometer

Inductively coupled plasma atomic emission spectroscopy.

## Pollution

Pollution is the introduction of contaminants into an environment that causes instability, disorder, harm or discomfort to the ecosystem i.e. physical systems or living organisms.

## Poplar

*Populus* is a genus of 25–35 species of deciduous flowering plants in the family Salicaceae, native to most of the Northern Hemisphere. English names variously applied to different species include poplar.

## Procambium

The primary meristem that gives rise to vascular tissue.

## Primary cell wall

A thin, flexible and extensible layer of the cell wall composed of cellulose, pectin and hemicellulose.

## Primary growth

Growth in vascular plants resulting from the production of primary tissues by a primary meristem. Elongation of the plant body is usually a consequence of primary growth.

## Potassium

Potassium is the chemical element with the symbol K, atomic number 19.

## Quercetin

Quercetin quercetin is a plant-derived flavonoid, specifically a flavonol, used as a nutritional supplement.

## Radyoacitivity

The level of radio- active substances in the eco- system and its components

## Rain water

The water sources used as a result of rain collection

## Regeneration

The ability to recreate lost or damaged tissues, organs and limbs.

## Respiration

The process in which nutrients are converted into useful energy in a cell.

## Rhizobia

Rhizobia are soil bacteria that fix nitrogen (diazotrophy) after becoming established inside root nodules of legumes (Fabaceae).

## Rhythm

Rhythm is the variation (fluctation) of the length and accentuation of a series of sounds or other events.

## Secretion

Secretion is the process of elaborating, releasing, and oozing chemicals from a cell, a secreted chemical substance or amount of substance.

## Secondary growth

In many vascular plants, secondary growth is the result of the activity of the vascular cambium.

## Secondary substances

Substances in vascular plants usualy formed from primary photosynthetic products. Often deposit in plant tissues

## Shikimic acid

Shikimic acid, more commonly known as its anionic form shikimate, is an important biochemical intermediate in plants and microorganisms.

## Shoot

Shoots are new plant growth, they can include stems, flowering stems with flower buds, leaves.

**Soil profile**

A vertical section of soil from the ground surface to the parent rock.

**Soil and land quality**

The determination of soil and eco- systems from the sens of its fertility

**Soil component**

It is composed of particles of broken rock that have been altered by chemical and environmental processes that include weathering and erosion.

**Soil contamination**

Soil contamination is caused by the presence of man-made chemicals or other alteration in the natural soil environment.

**Soil horizons**

A soil horizon is a specific layer in the land area which measures parallel to the soil surface and possesses physical characteristics which differ from the layers above and beneath.

**Soil micelle**

Soil unit consists of aluminum-silicate matrix and humus envelope

**Soil properties**

Physical and biological properties of soil and soil prophile

**Soil protection**

Some actions which support normal development of soil systems (normal pedo- genesis)

**Soil remediation**

The removal of harmful contaminants in soil

**Soil resources**

The spaces of fertile soils subjected to agro - ecological use

**Soil texture**

Soil texture is a soil property used to describe the relative proportion of different grain sizes of mineral particles in a soil.

**Solar energy**

Solar energy, radiant light and heat from the Sun, has been harnessed by humans since ancient times using a range of ever-evolving technologies.

**Species**

A taxonomic rank (the basic rank of biological classification).

**Space**

Space is the boundless, three-dimensional extent in which objects and events occur and have relative position and direction. Also- cosmos

**Space ecology**

Living organisms subjected to creation of eco-systems on some planets

**Starch**

Starch or amylum is a polysaccharide carbohydrate consisting of a large number of glucose units joined together by glycosidic bonds.

**Sulfur**

Sulfur or sulphur is the chemical element that has the atomic number 16. It is denoted with the symbol S. Essential element for plant life.

**Sumac**

Sumac is any one of approximately 250 species of flowering plants in the genus *Rhus* and related genera, in the family Anacardiaceae. Sumac is strong allelopathogene.

**Sustainable land management**

The treatments of land (landscape, agricultural use) without gene manipulations and chemical treatments of eco-cenosis

**Symbiosis**

The term symbiosis describes close and often long-term positive interactions between different biological species.

**Sustaining food production**

The production of organic food without gene transformations and chemical treatments (pesticides)

**Taxonomy**

Taxonomy is the practice and science of classification.

**T-DNA**

The transfer DNA (abbreviated T-DNA) is the transferred DNA of the tumor-inducing (Ti) plasmid of some species of bacteria such as *Agrobacterium tumafaciens* and *Agrobacterium rhizogenes*.

**Tobacco**

Tobacco is an agricultural product processed from the leaves of plants in the genus *Nicotiana*.

**Toxin**

A toxin is a poisonous substance produced by living cells or organisms.

**Transformed plants**

Plants which have plasmide modified DNA

**Tricarboxylic cycle**

The reverse Krebs cycle (also known as the reverse tricarboxylic acid cycle, the reverse TCA cycle, or the reverse citric acid cycle) is a sequence of chemical reactions that are used by some bacteria to produce carbon compounds from carbon dioxide and water during respiration.

**Tryptophan**

Tryptophan is one of the 20 standard amino acids, as well as an essential amino acid in living organism.

**Ultraviolet**

Ultraviolet (UV) light is electromagnetic radiation with a wavelength shorter than that of visible light, but longer than x-rays, in the range 10 nm to 400 nm.

**Vitamins**

A vitamin is an organic compound required as a nutrient in tiny amounts by an organism.

**Waste water**

Wastewater is any water that has been adversely affected in quality by anthropogenic influence. It comprises liquid waste discharged by domestic residences, commercial properties,

**Water treatment**

Water treatment describes those processes used to make water more acceptable for a desired end-use

**Water cycle**

The water cycle, also known as the hydrologic cycle, describes the continuous movement of water on, above and below the surface of the earth.

**Weed**

A weed is a plant that someone thinks is bad, because it is growing in the wrong place.

**Wetland construction**

A constructed wetland or wet park is an artificial marsh or swamp, created for anthropogenic discharge such as wastewater, storm water runoff or sewage treatment, and as habitat for wildlife, or for land reclamation after mining or other disturbance.

**Willow**

Willows, sallow's, and osiers form the genus *Salix*, around 400 species of deciduous trees and shrubs, found primarily on moist soils in cold and temperate regions of the Northern Hemisphere.

**Wood**

Wood is an organic material; in the strict sense wood is produced as secondary xylem in the stems of trees (and other woody plants).

**Wood fiber**

Wood fibers are usually cellulosic elements that are extracted from trees, straw, bamboo, cotton seed, hemp, sugarcane and other sources.

**Woody plants**

A woody plant is a plant that uses wood as a structural tissue. They are typically perennial plants that have their stems and larger roots reinforced with wood produced adjacent to the vascular tissues:

**Xanthomutant**

Plant mutant which lost green color (chlorophyll content).

## Xylem

In vascular plants, xylem is one of the two types of transport tissue, phloem being the other. Xylem is plant conducting tissue.

## Zinc

Zinc also known as spelter, is a metallic chemical element; it has the symbol Zn and atomic number 30. Zn is micro-element in plant mineral nutrition

# INDEX

## A

2,4-D –see 2,4-Dichlorophenoxyacetic acid
ABA -see Abscisic acid
Abscisic acid (ABA) 41, 43, 51-53, 56-61, 65, 66, 99, 109
Absorption 1, 85, 99, 100
Actinomycetes 23, 27, 33, 38, 99
Adaptation 97, 99
Agricultural land 8, 11, 13-15, 32, 96, 99
Agrobacterium tumafaciens 49, 50, 57, 99,116
Agronomic crops 45, 99
Agrosol 17, 99
Albino 59, 60, 61, 63, 99
Alkaloid 64, 66, 77, 99
Alfisol 16, 18, 88 92, 93, 99
Allelopathogenes 100
Allelopathy 45, 47, 64, 76, 100
Amino acids 32, 40, 53, 64, 66, 86, 89, 90, 100, 117
Ammonium 32, 33, 63, 100, 112
Anthocyanin 43-45, 53, 60, 61, 64, 65, 77, 100
Anthranilic acid 54, 100
Atomic absoption spectrophometry 100
Atomic emission spectrophometry 100
Auxin 42, 46-48, 50-53, 56-61, 67-70, 75, 79, 85, 88, 100, 101, 108, 110
Auxin oxidase 47, 51, 59, 101

## B

BA- seeBenzyladenine
Bacillus 23, 24, 26, 27, 34, 101
Bacteria 20-28, 31-39, 49, 56, 84-86, 89, 90, 100
Bacterial count 23,27,37,38
Bark 40, 47, 101
Benzyladenine (BA) 49, 50, 51, 57, 58, 101
Bioflora 24, 28, 101
Biological activity 19, 22, 26, 28, 29, 37, 38, 39, 44
Biological test 95, 101
Biomass 1-12, 22, 25, 26, 32, 40, 42, 77, 78, 86, 87, 94
Biomass accumulation 40, 77, 101
Biomass production 1, 2, 5, 6, 9, 12, 86, 94, 95, 101
Bio-minesoil 19,101
Biosphere 1, 13, 17, 21, 36, 78, 101, 102
Biosynthesis 32, 40, 42, 44, 47, 52, 53, 75, 76, 78, 89, 90, 101, 102, 106, 109
Biotechnological models 85, 91, 97, 101
Biotest 50, 57, 63, 102
Boron 37, 102, 110
Bud 47, 62, 64, 65, 68, 70-73, 99, 102, 109, 114

## C

C-3 plants 40, 102

www.ingramcontent.com/pod-product-compliance
Lightning Source LLC
Chambersburg PA
CBHW041716210326

41598CB00007B/671